The Martin P5M

Marlin

by Bruce D. Barth

Published by Pacific Aero Press
P.O. Box 20092
Castro Valley, California 94546-8092

ISBN 0-9636332-1-X
First Printing May 1994
Printed in the United States of America

The Martin P5M "Marlin"

Layout by ***LazerTouch***
San Mateo, California

Front Cover:

Squadrons often flew both the P5M-1 and P5M-2 *Marlins* simultaneously as shown here by VP-44 in 1958. (U.S. Navy photo via Robert L. Lawson)

✪

Patrol Squadron Forty-Four was the first squadron to receive the P5M-1 *Marlin* in 1952. Displayed here is their squadron insignia from that era. (Author's Collection)

Back Cover:

Shown on November 6, 1967 is an SP-5B from VP-40, departing the San Diego seadrome on the last operational flight by a U.S. Navy seaplane squadron. At the controls of this last flight is Cdr. Joseph P. Smolinski, Jr., the commanding officer of the last naval seaplane squadron to fly the Martin *Marlin*. (U.S. Navy photo via John Fry)

✪

The squadron painting shown was commissioned in 1989 by Bruce D. Barth and depicts the history of VP-40's squadron insignia throughout their long association with Martin seaplanes. (Author)

Dedication

The added dimension of a waterborne environment required additional demands of seaplane crews. These included occasional dangerously high sea-states for takeoffs and landings, maintenance of watertight integrity in a highly corrosive saltwater environment and the ennui of taxiing around for hours awaiting an opportunity for beaching the aircraft after a long arduous patrol.

A hearty "well done" to all those seaplane crewman and their supportive families who extracted every ounce of the seaplane's capability to accomplish her ASW and other assigned missions. They earned high respect for their unselfish contribution to the eventual ending of the "Cold War."

Captain Joseph P. Smolinski, Jr.
U.S. Navy (Ret.)

Acknowledgments

My professional interest in aviation and fond memories of my association with the Martin SP-5B have been the thrust behind my compelling desire to complete this book. However, although a portion of the material presented herein is based on personal experience, I cannot fully accept the credit for the research necessary to complete this publication. Furthermore, credit for the completion of this book must be shared with the organizations that supplied materials, and the countless individuals who donated their time, shared personal experiences and materials and provided me with encouragement over the years. Listed partially, in no particular order, they are;

Capt. Joseph P. Smolinski, Jr., USN Ret.; Capt. Richard A. Hoffman, USN Ret.; Capt. John R. Johnson, USN Ret.; Capt. Irving G. Cockroft, USN Ret.; Capt. Vadym V. Utgoff, USN Ret.; Capt. Frank J. Schneider, USN Ret.; Capt. Jack Sullivan, USNR Ret.; Capt. Bruce H. Handler, USNR Ret.; Cdr. Louis M. Guerrieri, USN Ret.; LCdr. Donald R. McGarrigle, USNR Ret.; Richard C. DeLaMater; Nicholas A. Veronico; Richard H. Benbough; A. Kevin Grantham; Grace D. Schmidt; Nadine S. Barth; Debbie Arnold; Clay Jansson; Bob Smith; Ed Davies; Stan Piet; Glenn L. Martin Aviation Museum; Mariner / Marlin Association; Shin Meiwa Industries, Ltd.; Naval Aviation History Division; Naval Aviation News; Patrol Squadron Forty; National Museum of Naval Aviation; National Air & Space Museum; VP-40 Reunion Group; VP-45 Reunion Group.

The Martin P5M *Marlin*

the end of an era in naval seaplane aviation

Table of Contents

The Martin P5M *Marlin*

the end of an era in naval seaplane aviation

List of Illustrations

The Martin P5M *Marlin*

the end of an era in naval seaplane aviation

Introduction

This publication is intended to provide a comprehensive accounting of the U.S. Navy's last production seaplane, from its post-war development in the late 1940s to its untimely demise in the mid-1960s. Also included, is a complete history of the Navy's last seaplane squadron, Patrol Squadron Forty.

The Navy purchased its first seaplane on May 8, 1911. Accepted two months later on July 6, 1911, was the Curtis A-1 *hydroaeroplane*. By 1918, the Navy's complement of seaplanes grew to outnumber their land based counterparts by nearly nine to one. In the decades that followed, the inventory of flying boats continued to increase and became an invaluable part of naval aviation. By the 1940s, the seaplane era had reached a pinnacle with the Navy, with designs such as the Martin PBM *Mariner* and *Mars*, and the Consolidated PBY *Catalina.* They proved themselves a legend during World War Two. All saw action in almost every theater of the war. Each seaplane played a major role in anti-submarine warfare (ASW), aerial mining, bombing, ocean surveillance and air-sea rescue operations throughout the world.

Advancements in seaplane design took on many changes over the years. Recognized as a venerable threat against submarine warfare, these designs continued to improve, and in 1952, the Navy took delivery of the Martin P5M *Marlin.* A manufacturer of seaplanes since 1909, Martin had designed and produced a succession of rugged and reliable flying boats for Naval and Coast Guard use. The introduction of the Martin *Marlin* was a fresh new addition to the military's fleet of aging seaplanes. Born out of war-time experience with the successful PBM *Mariner*, the P5M was the first seaplane designed and built primarily for ASW use by the Navy.

Conceived in 1946 and after five years of development, the P5M arrived too late to prove itself in the Korean War; however, the P5M soon found a home among a growing number of post-war patrol squadrons throughout the United States, Pacific and Caribbean. Incorporating the latest concepts in seaplane technology, the P5M played a pioneering role as a development tool for advanced aero-hydrodynamic programs for future designs.

Before production of the newly designed P5M-1 model concluded, Martin introduced an improved version designated the P5M-2 in 1954. Although future development and design of naval seaplanes continued into the late 1950s, the Martin P5M-2 *Marlin* was the last production design accepted by the Navy. After nearly fifty-six years, the Navy's long-running association with flying boats was destined to end where it all began back in 1911 at North Island, San Diego, California. Without a suitable replacement for the P5M, the future of the seaplane in a military role was just marking time to progress.

In August 1958, Lockheed Corporation of California won the competition for the next generation of maritime patrol aircraft with the introduction of the P-3A *Orion*. Derived from the company's commercial transport version of the four-engine *Electra*, the Lockheed designed *Orion* would ultimately replace all patrol aircraft in naval service. Placing an initial order for seven of the new turbo-prop aircraft in 1960, the Navy's lengthy career flying the seaplane was soon to draw to a close. Delivery of the P-3A commenced in August 1962 and soon after the number of patrol squadrons flying the Martin P5M began to dwindle.

By 1965 only three seaplane squadrons remained in operation, VP-40, 48 and 50. Redesignated the SP-5B in 1962, the *Marlin* flying boat would play one last significant role with the Navy before its retirement, a final tribute to the end of an era in naval seaplane aviation.

Commencing in 1965, the Navy's three remaining seaplane squadrons rotated continuous deployments to the Pacific in support of the Vietnam conflict, flying support for *Operation Market Time* off the coast of Vietnam. This joint Navy effort, between surface and air forces, provided ocean surveillance for communist cargo shipping and ASW operations in the Tonkin Gulf and South China Sea. By 1967, only one seaplane squadron remained in operation.

To support the Korean War, VP-40 was re-formed in January 1951 and was the second squadron to receive the P5M *Marlin* in April 1953, and the last squadron to fly them in November 1967.

Following one final tour in Vietnam with VP-40, the last of the Navy's *Marlin* seaplanes were retired at NAS North Island on November 6, 1967, officially ending the career of the flying boat with the U.S. Navy. For the City of San Diego, which saw so many aviation pioneers come and go over the years, there seemed to be a particularly strong sense of loss over the seaplane which had been a part of its aviation history at North Island for over fifty-six years. For the last squadron to fly the *Marlin*, it was not only an end of an era, but the end of nearly three decades of flying Martin seaplanes.

Marlin

P5M SEAPLANES

Preface
The PBM "Mariner"
(1937 to 1949)

As early as 1937, the U.S. Navy was considering designs for a larger, stronger flying boat to replace the PBY *Catalina*. While competing companies scrambled to enter the competition with four-engine designs, the Glenn L. Martin Aircraft Company, Baltimore, Maryland, proposed a twin-engine design designated the Model 162.

The Model 162 was a clean design from the onset. Configured with an unusual 118 foot gull wing, deep hull, retractable wing-tip floats, nacelle enclosed bomb bays and Martin-designed gun turrets, the Model 162 flying boat looked very promising. But the Martin Aircraft Company, serious about the proposed design and the competition, decided to further prove their own theory for the new flying boat by building a scale model prototype. An industry first, Martin constructed a three-eighths scale model designated the 162-A. Power for the manned flying model came from a single Martin 4-333 inverted four-cylinder air-cooled aviation engine. Mounted amid-ships, the small engine, designed by Louis Chevrolet, supplied power to both propellers through a series of belts. Successfully tested, the flying model not only proved the design, but provided Martin with invaluable hydrodynamic experience for the new concept in flying boats. On June 30, 1937, the Navy awarded Martin a contract to build a single full-scale prototype under the designation XPBM-1.

Based on the impressive performance of the scale model, the Navy placed an advanced order for twenty production models and a single XPBM-2 on December 18, 1937, fourteen months before the prototype was test flown. Completed in late 1938, the prototype XPBM-1 made its maiden flight on February 18, 1939. Assigned the production designation PBM-1 *Mariner*, the model was accepted by the Navy on September 7, 1939. The first production lot of twenty PBM-1s were distributed among two newly commissioned squadrons, VP-55 and VP-56, in 1940.

Ideally suited for combat operations, the PBM quickly proved itself with the Navy as a rugged replacement for the PBY and additional orders were placed for an improved version. By late 1942, the PBM-3 was introduced. Offering significant improvements over that of the original model, the PBM-3 was manufactured in several versions including 3C, 3D, 3R and 3S. The last model in the PBM series was the PBM-5. Introduced in the summer of 1944, the dash-five was the only model in the series that was also pro-

The 1,700 pound, three-eighths scale Model 162-A is shown resting in its beaching cradle outside the Martin hanger in early 1938. Flown by a single pilot, this aluminum and spruce prototype was capable of speeds to 120 miles per hour. It is now owned by the Smithsonian Institution and was restored in 1992 for display at the Baltimore Museum of Industry.
(Martin Aircraft Company)

duced as an amphibian (PBM-5A). Production of the PBM series flying boat concluded in March 1949, with a total of 1,368 aircraft produced.

The PBM-5 and 5A continued to operate with both Navy and U.S. Coast Guard squadrons throughout the Korean War, receiving continuous updates for improved performance, ASW capabilities and ordnance. These aircraft were designated -5S and -5S2. Following the end of the Korean War in mid-1953, many *Mariner* squadrons found themselves making a transition to newer patrol aircraft. By 1956, the number of fleet squadrons flying the *Mariner* had been reduced to one. On July 31, 1956, VP-50 retired the PBM-5S2 at NAS Alameda, California and made the transition to the Navy's latest patrol aircraft, the Martin P5M-2 *Marlin*.

The first XPBM-1 *Mariner* prototype, BuNo 0796, is shown here in 1939. This prototype was fitted with retractable wing-tip floats and a straight horizontal stabilizer. On production models, dihedral was added to the stabilizer to match that of the gull wing.
(Warren M. Bodie Collection)

The XPBM-1 is shown in flight with wing-tip floats retracted. This 1939 photo, taken during early flight testing, clearly shows the characteristic high bow chine and deep single-step hull used on the *Mariner* series.
(Warren M. Bodie Collection)

Pictured in this 1942 photo is an early PBM-3 on takeoff. Produced in several versions, the PBM-3 introduced fixed wing-tip floats and was the predominant model in the *Mariner* series. Notice the added dihedral in the horizontal stabilizer.
(Warren M. Bodie Collection)

The PBM was also used by the USCG. This 1948 photo demonstrates the use of JATO on a PBM-5G at Salem, Ma. Note the early radome.
(Warren M. Bodie Collection)

A PBM-5A amphibian on takeoff. The aircraft pictured is configured with three Martin gun turrets and a late model radome over the cockpit. Production models used R-2800 engines, turning four-bladed Curtis Electric propellers. Only forty PBM-5A amphibians were produced, the last in 1949.
(Warren M. Bodie Collection)

Shown at NAS Norfolk is the last revision in the PBM-5 series, the PBM-5S2. This model continued to operate with VP-50 until 1956.
(Warren M. Bodie Collection)

The Martin P5M
Marlin

Chapter One

The P5M "Marlin" Model 237 XP5M-1

As early as 1944, Martin produced several conceptual designs for advanced medium-range flying boats to succeed the PBM-5. Many of these designs included a longer, slimmer, single step hull, refined armament and conventional tail assembly. By 1945, Martin had completed a design for the next generation PBM, the PBM-6. Nearly eight feet longer and five feet higher than its predecessor, the twin-engine PBM-6 design employed a single step hull, extending the length of the fuselage, a conventional tail assembly and stronger wing-tip floats. Although not immediately accepted by the Navy, this design concept was of significant importance. Approximately one year later, Martin was approached to design a new generation medium-range patrol flying boat and, it would be the PBM-6 concept that would be used as the foundation for the project.

On June 26, 1946, Martin was awarded a contract to develop and produce a prototype flying boat to replace the famous *Mariner* which had been in service since 1939. Based on the PBM-6 design, Martin initiated a project designated, Model 237. Development of the prototype started with a standard production PBM-5, BuNo 98616, the last aircraft manufactured in the model series. Retaining the standard 118 foot *Mariner* wing and floats, and extending the long after body of the fuselage some eleven feet, including a new tail assembly, Martin engineers added a pair of 2,700 horsepower Curtiss-Wright R-3350-26W engines, thus forming the basic design for the successor to the PBM *Mariner* series flying boat.

Designated XP5M-1, the highly modified prototype differed somewhat from the original PBM-6 concept. Longer in length, with a slightly taller tail assembly, the prototype retained the original dash-five wing tip floats while several new designs were being evaluated. Martin designers also considered an amphibious version, using a refined retractable landing gear system similar to that used on the PBM-5A; however, like the earlier model (PBM-5A), the concept was abandoned early on due to the excessive weight of the landing gear assemblies.

The all new and more efficient single step "V" hull, with long after body and greater length-to-beam ratio (8.5 to 1), greatly reduced the tendency to porpoise (longitudinal instability) on rough water, a problem that had plagued earlier designs. Another unique feature was the use of Martin designed *hydroflaps,* incorporated to improve handling during waterborne operation. Controlled like conventional toe brakes on the rudder pedals, these hydraulically actuated flaps had an operating limit of sixty-five degrees and reduced the normal turn radius from 800 feet (PBM) to less than 300 feet. Located under the water line below the tail, they could be deployed together, acting to slow the aircraft or separately to provide directional control while taxiing.

This concept drawing of the XP5M-1 was based on a 1945 design for the PBM-6.
(Martin Aircraft Company)

When used in conjunction with the Curtis Electric reversible pitch propellers, the XP5M-1 became extremely maneuverable on the water—unlike its predecessor which relied on the use of a canvas *sea anchor* for directional control.

The use of conventional defensive armament was also considered in the new design to include radar-controlled, power operated twin twenty-millimeter turrets in the nose and tail positions, and a manned power-driven turret in the dorsal position aft of the wing. Each engine nacelle was also extended to allow for a large bomb bay capable of carrying multiple stores for offensive use. The tail was completely redesigned, deleting the twin rudders of the older PBM in favor of a tall, single vertical-fin/rudder and conventional low mounted horizontal stabilizer. Completed by early 1948, the prototype made its first flight on May 30, 1948 from the Martin factory in Baltimore.

The XP5M-1 prototype is shown here at NATC, Baltimore in 1950. This aircraft was based on a standard production model PBM-5, BuNo 98616.
(Warren M. Bodie Collection)

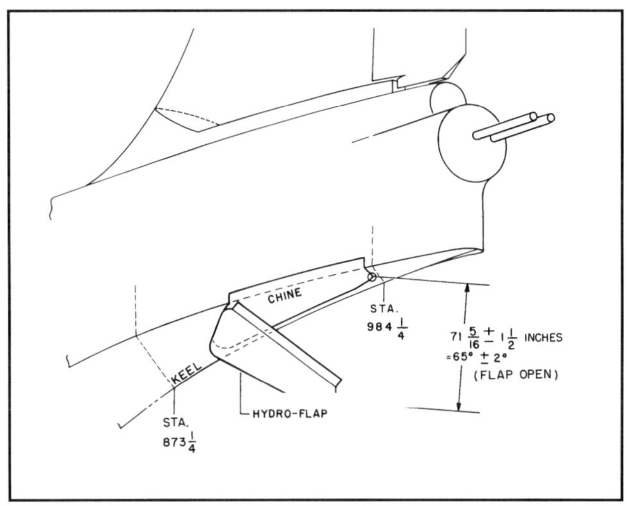

Pioneered on the *Marlin,* the Martin designed *hydroflap* was used as a means of improving the maneuverability of the aircraft on the water.
(Martin AN 01-35EJB-2)

This silhouette drawing compares the PBM-5 to the new XP5M-1.
(Martin Aircraft Company)

Artist concept drawing of an early P5M-1, circa 1949.
(Martin Aircraft Company)

P5M-1

After nearly eighteen months of extensive testing, evaluation and modification, a refined version of the prototype was accepted by the Navy in late 1949. Production of the new aircraft, now designated the P5M-1 *Marlin,* commenced with an initial order for twenty-five, in July 1950. The initial flight of the first production P5M-1 took place on June 21, 1951 under the direction of O. E. (Pat) Tibbs, Director of Flight for Martin.

The Navy took delivery of their first *Marlins* in 1952. A ceremony was held on April 23 by VP-44, NAS Norfolk, Virginia, to formally receive the first production models of the new aircraft. Typical of such events, the squadron's first *Marlin* was officially christened by the wife of the commanding officer. Commanding Officer Cdr. Francis J. Grisko, USN, presided at the ceremony.

Almost exactly one year later the *Marlin* received a similar welcoming at its first West Coast squadron, VP-40. Based at NAS North Island, VP-40 received their complement of the new Martin flying boat from FASRON-110 (Fleet Air Service Squadron), following their return from deployment to the Pacific. The event was marked by an identical ceremony on April 30, 1953 under the direction of Cdr. Joseph M. Kellam, USN, commanding officer.

Emerging as a completely new aircraft, the production version differed considerably from that of the prototype. An obvious difference was the deletion of most of the prototype's defensive armament. The dorsal turret was removed in favor of a manned, twin, twenty-millimeter cannon turret (Aero 11D-1) mounted in the tail. These were aimed and controlled by a gunner positioned just forward and above the turret. Sighting was accomplished using a Mark 18 Model 6 lead-computing gunsight through a domed Plexiglas window just above the turret. The nose turret was replaced by a powerful search radar (AN/APS-44A), distinguishable by a large protruding radome. In addition, the cockpit was elevated to improve overall visibility and new concept wing-tip floats were installed. Using a single streamlined pylon, braced with an inverted "V" strut, the new stabilizing float design proved itself a much stronger structure over that of the earlier PBM which had a propensity to break away prematurely.

Uprated to the more powerful R-3350-30WA Turbo-Compound engines, the production version saw an increase in horsepower to 3,250 each over the earlier pre-production 2,700 horsepower engines on the XP5M-1. Equipped with an alcohol-water injection system, two-speed supercharger and power recovery turbines (PRT), the eighteen cylinder engines were the second largest in the radial engine class at the time.

The first production P5M-1, BuNo 124910, shown during initial flight tests in 1951. Notice the long after-body hull design, extending the length of the aircraft.
(U.S. Navy photo)

Shown is a Patrol Squadron Forty ceremony at which the first West Coast *Marlins* were received, at NAS North Island on April 30, 1953.
(U.S. Navy photo via R. E. Dickinson)

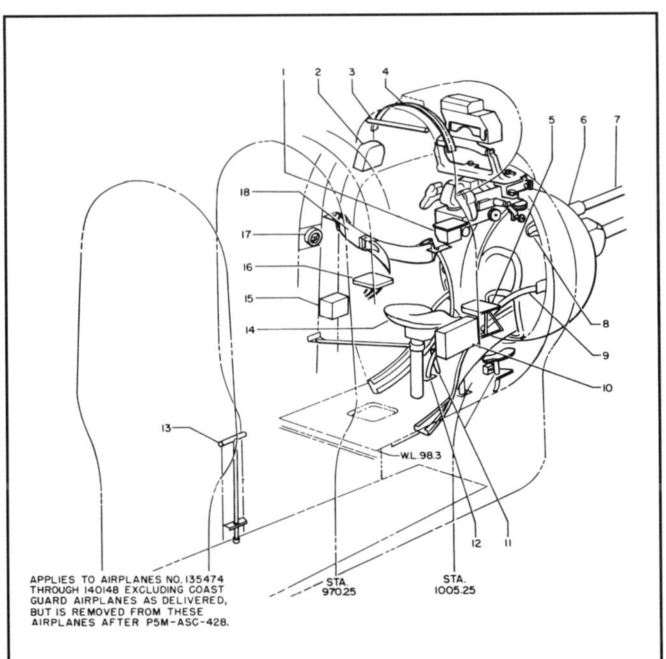

This illustration shows the components of the *Marlin's* manned, twin twenty-millimeter tail turret. As part of the Modification Program, by 1958 the tail turret had been deleted on most models and replaced by additional ASW equipment.
(Martin AN 01-35EJB-2)

Operated by exhaust gas through liquid couplings, the three power recovery turbines returned between sixteen and twenty percent power to the crankshaft. A unique feature of the turbo-compound engine, this was a self-sustaining means of increasing horsepower without a loss of operating efficiency.

The production models were additionally fitted with improved Hamilton Standard *hydromatic* four-blade propellers which replaced the prototype's Curtis Electric versions. At fifteen feet in diameter, these aluminum alloy propellers were both reversible and full-feathering and incorporated electric de-icing blankets on the leading edge of each blade. Flight control of the large aircraft was improved greatly with the introduction of hydraulically actuated spoiler ailerons and power-assisted controls.

Equipped with a basic complement of seven crew members; two pilots, a navigator, radioman, two electronics operators and a tail gunner, the cavernous fuselage of the *Marlin* was comprised of fourteen compartments separated by three internal watertight doors. The larger internal dimensions offered improved seating and other significant improvements in crew comfort over the older PBM design. Nonetheless, in some cases the new *Marlin* lacked some of the more positive features of its predecessor, such as electric bomb hoists, anchor windless and a relatively sound-proofed bunk area. These deficiencies were not corrected in subsequent revisions in the P5M series.

The two large-engine nacelles housed symmetrical bomb bays on the underside of the center wing and with the addition of underwing racks, provided for up to 8,000 pounds of ordnance. Typical internal stores included aerial mines, bombs, torpedoes, depth charges or auxiliary fuel tanks. Underwing racks were provided in two configurations, allowing for the use of bombs and or aerial rockets, including three-inch SCAR (Solid Core, non-explosive ASW) and five-inch HVAR (High Velocity, explosive head) rockets. In all, thirty-five combinations of internal and external stores could be carried by the *Marlin.*

The two 575 gallon auxiliary fuel tanks, which could be electrically jettisoned if necessary, greatly extended the range of the *Marlin* and were often used on long distance ferry missions, including trans-Pacific flights. Additional offensive submarine detection stores included both large and small sonobuoys, MK-7 MOD. 1 retro-marine markers, MK-6 float lights and MK-5 parachute flares.

Search operations were further enhanced by use of a new, 70 million candle power carbon-arc search light. Remotely controlled from the co-pilot position, the AN/AVQ-2 searchlight was mounted via a pylon on the underside of the right wing tip and could be jettisoned electrically if necessary. Due to the extreme heat generated by the carbon-arc system, operation of the new search light was limited to a thirty second on, four and one half minute off, duty cycle.

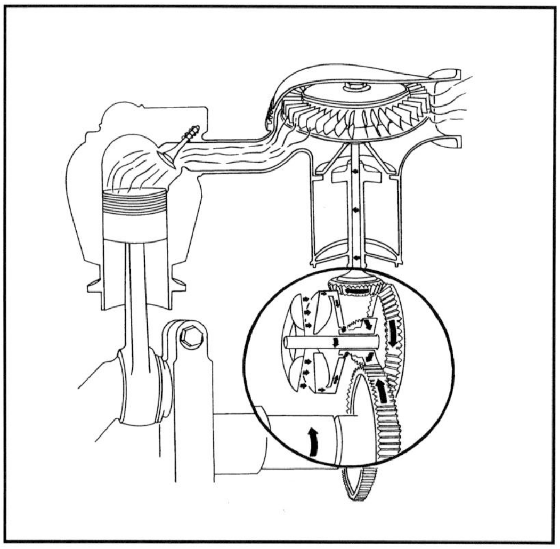

These power recovery turbines, used on the R-3350, returned up to twenty percent power to the crankshaft through liquid couplings.
(Martin AN 01-35EJB-2)

This cut-a-way photo of BuNo 124910 shows the main compartments of the P5M-1. In all, the *Marlin* had fourteen compartments, separated by three watertight doors, on two levels. This nose mounted search radar was the largest of its kind. (Martin Aircraft Company)

Although seemingly short in duration, the operational cycle was more than adequate under all but the most demanding circumstances.

With a normal gross operating weight of 73,000 pounds, the P5M-1 was the largest seaplane in its class at the time. The huge R-3350 radial engines, developing 3,250 horsepower each, were more than adequate for takeoff under most operating conditions; however, when operating under heavy sea conditions, or excessive weight loads, the *Marlin* required the use of a Jet Assist Take Off system, known as JATO. Pioneered in the mid-1940s, these electrically fired, solid-fuel booster rockets provided up to 1,000 pounds of thrust each for approximately fifteen seconds. The P5M had provisions for mounting four JATO rockets, two of which attached to each waist hatch on each side of the aircraft.

Once attaining a predetermined takeoff speed based on conditions, the 4,000 pounds of thrust provided by the booster rockets would accelerate the aircraft off the water. When airborne, the spent rocket motors were then jettisoned into the sea by use of a manual release system on each hatch.

Designed primarily as an all-weather ASW patrol aircraft, the P5M-1 incorporated a full range of anti-icing equipment to include automated, pneumatic, boot de-icing on the leading edge of control surfaces; electrically heated propeller blades; windshield protection, including wipers; alternate warm-air carburetor heat, and protection of oil cooler scoops.

Typical performance characteristics of the *Marlin* included an operating range of nearly 3,000 miles at a cruising speed of 157 miles per hour (136 knots), with a maximum airspeed of 269 miles per hour (234 knots) at sea level.

In its ASW role, the non-pressurized P5M typically operated at an altitude of less than 10,000 feet and was equipped with a manual oxygen system for use up to its maximum service ceiling of 21,000 feet. General requirements for the use of oxygen aboard military aircraft was specified at above 10,000 feet by day and 5,000 feet at night. The pilot and co-pilot positions each contained a fixed oxygen system consisting of two 514 cubic inch cylinders, interconnected, with diluter demand regulators that could be used independently as necessary. Ten additional 295 cubic inch portable oxygen units were also carried aboard for use by crew personnel and as replacements. As a general rule for added safety, twice as many portable units were carried than actual number of crew personnel aboard, excluding the pilot and co-pilot.

GENERAL ARRANGEMENT

Before P5M-ASC-439

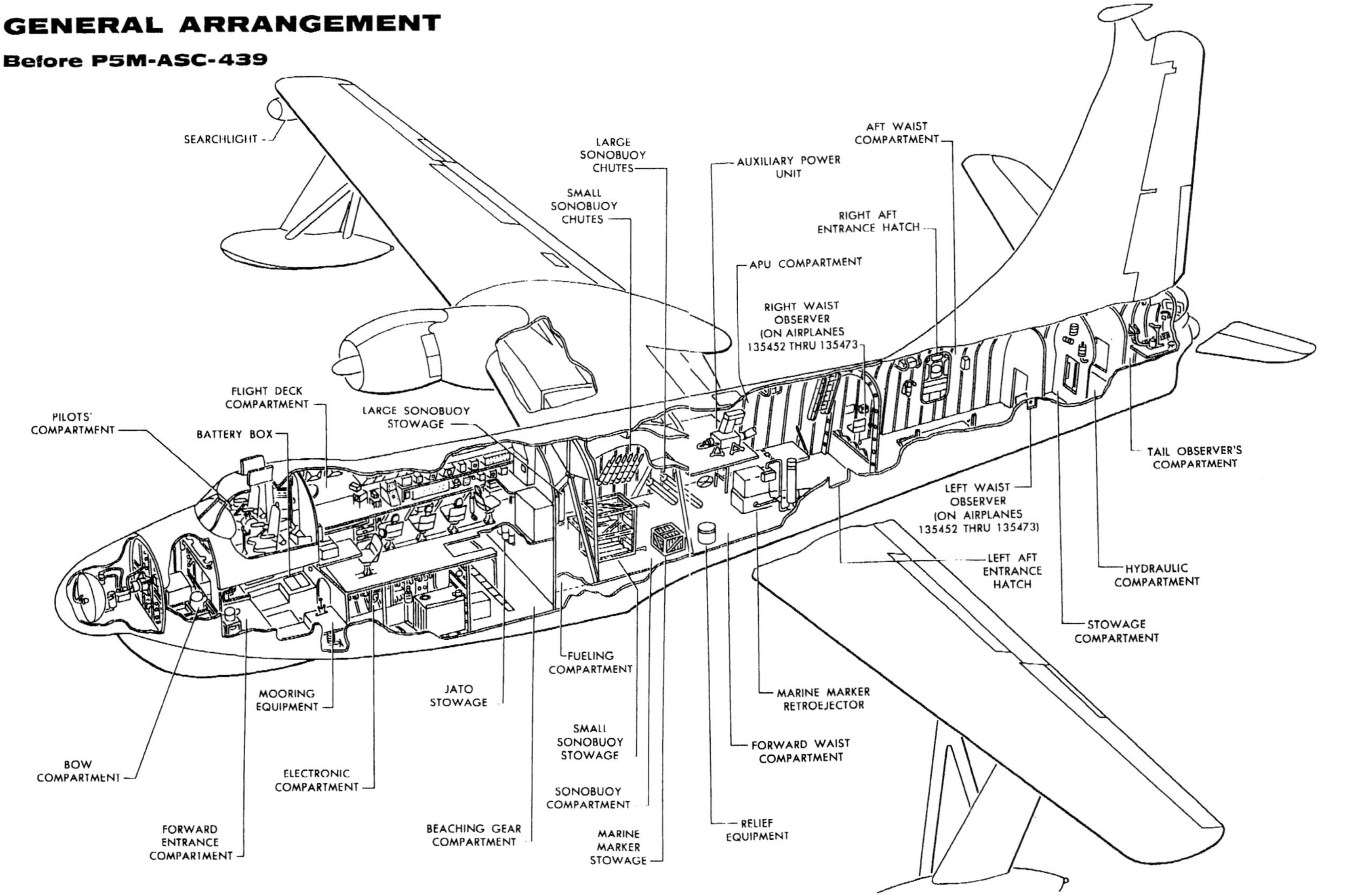

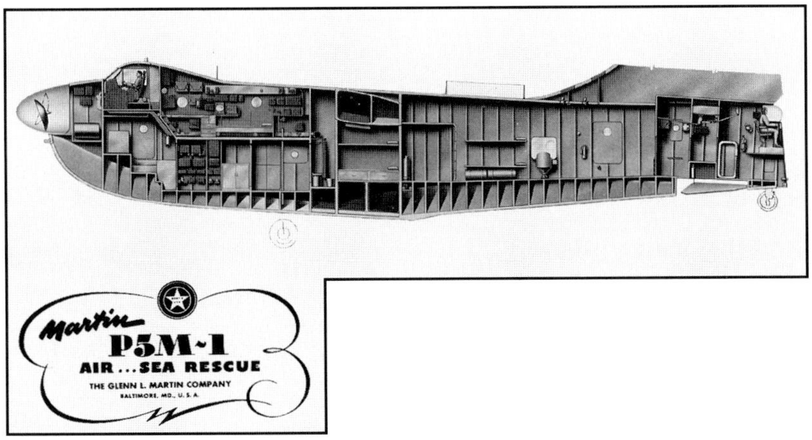

Designed for use by the USCG, the air-sea rescue version of the *Marlin* was manufactured without the Navy's ASW electronics and armament and was available with a smaller search radar, the AN/APS-33. (Martin Aircraft Company)

The electrical power necessary to support the *Marlin's* large electronic suite was derived from a complex electrical system managed and controlled through a master electrical panel located at the Radio Operator's position. Comprised of both A-C (Alternating Current) and D-C (Direct Current) generating systems, conventional batteries and power converters, the electrical system was connected through a main power distribution center located aft of the main electronics compartment.

The A-C power system was a 115/200 volt, three-phase, 400 cycle system, comprised of three, 40 KVA a-c generators connected in a four-wire grounded neutral system. Two of the three a-c generators were driven by the main engines through *Sundstrand* constant speed drive transmission units. The third generator was connected directly to the Auxiliary Power Unit (APU). Powered by a small AirResearch gas turbine (Model GPT 70-10), the APU 40 KVA generator was used as a source of power when the aircraft was moored or during waterborne operations.

Power for the aircraft's 28 volt D-C system was supplied by two engine-driven d-c generators and one 200 ampere regulated converter. Each d-c generator had a continuous rating of 200 amps at 30 volts, controlled by a carbon-pile voltage regulator. The 200 ampere regulated converter consisted of a transformer-rectifier combination used to change the 115/200 volt, three-phase power input to 28 volt d-c output. The aircraft also contained two, 36 ampere, 24 volt, lead-acid batteries located in the battery compartment on the right side of the bow compartment.

The hydraulic system on the *Marlin* consisted of two primary 3,000 psi (pounds per square inch) systems and one emergency back-up system. The main system was powered by two, electrically driven, variable-volume pumps, controlling the spoiler ailerons, wing flaps, hydroflaps and bomb-bay doors. The rudder and elevator control system had a separate reservoir, two electric pumps, a filter, and relief valves. Each pump was powered by a different electrical bus to ensure maximum sys-

The *Marlin* was capable of mounting four JATO booster rockets, as demonstrated by this P5M-1 at Strawberry Point in March 1952. (Martin Aircraft Company)

LANDING PATTERN

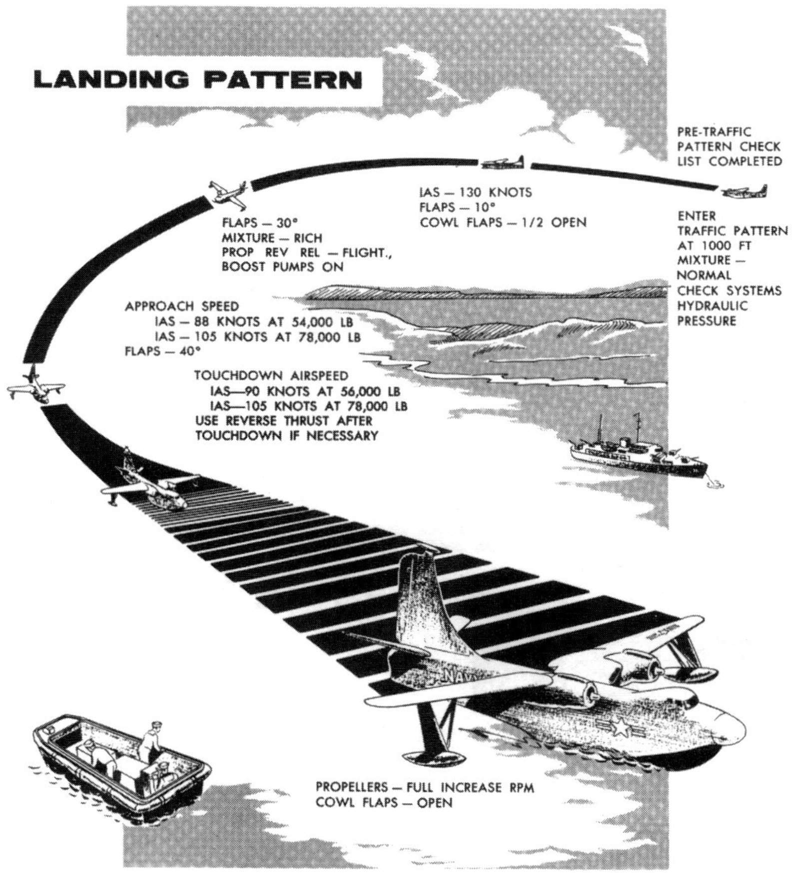

MOORING PROCEDURE

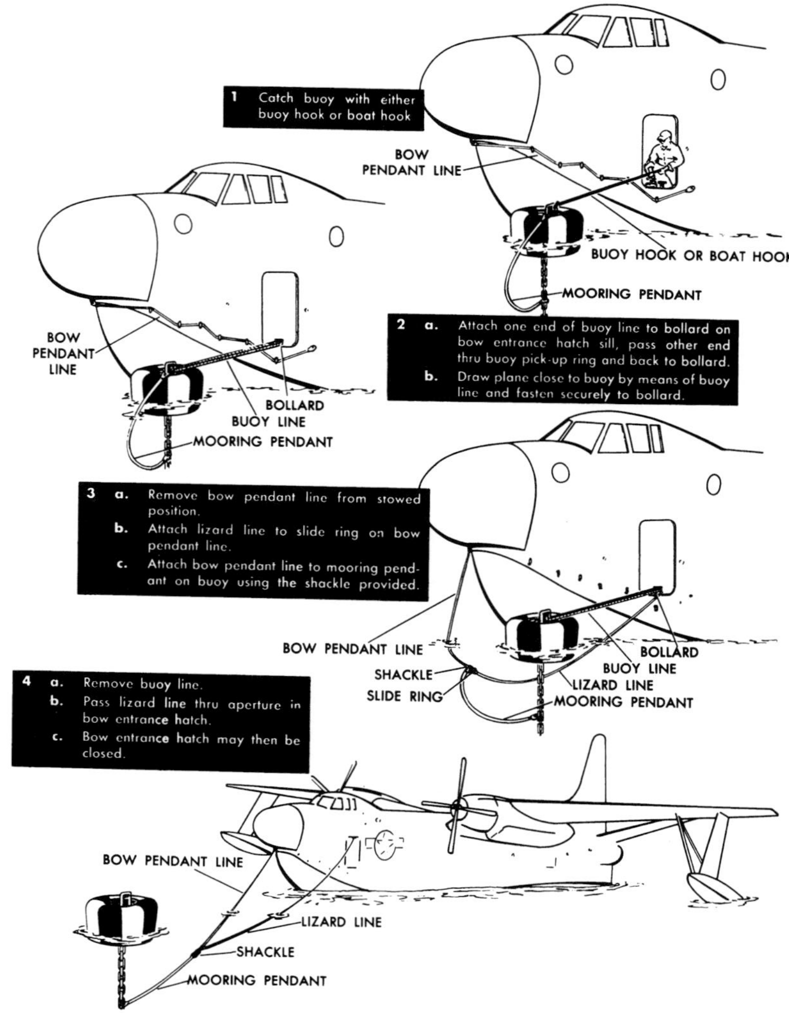

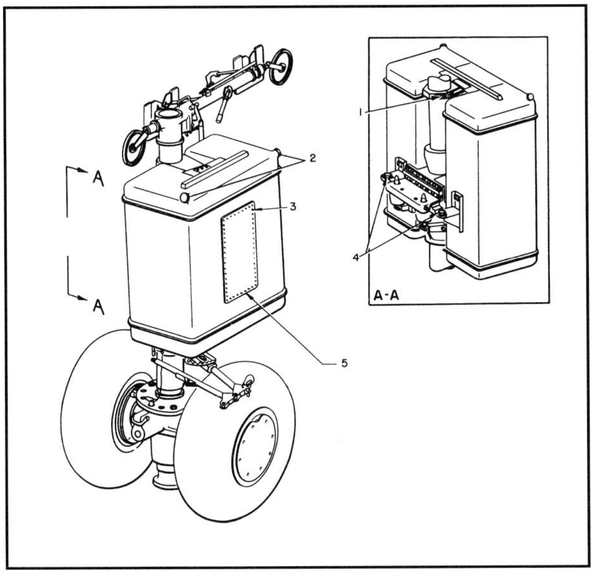

Each main beaching gear was capable of supporting 78,000 pounds and was equipped with a flotation tank and a hand brake assembly.
(Martin AN 01-35EJB-2)

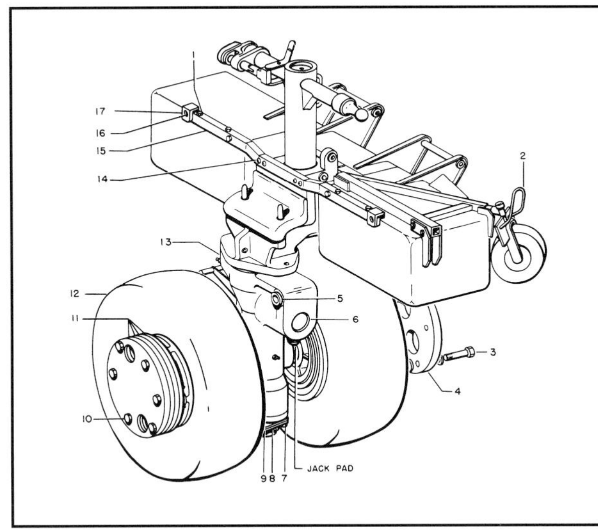

Mounting of the tail beaching gear was accomplished through the combined effort of a crewman inside the aircraft and beaching personnel in the water. It was accessed through a small hatch in the tail. (Martin AN 01-35EJB-2)

tem reliability and separately each had sufficient capacity to operate both rudder and elevator surfaces during normal maneuvers. A separate emergency back-up system was also provided through use of a manually operated hand pump. This system was designed to provide hydraulic pressure for operating the flaps and bomb bay doors in the event of electrical failure.

It should be noted that early models of the P5M had two additional independent hydraulic systems. There was one each for the twenty-millimeter tail turret and the AN/APS-44A search radar. These were deleted on later production models and on those that were upgraded during the modification program.

With an operating range of nearly 3,000 miles and up to fourteen hours endurance, the *Marlin's* multi-tank fuel system contained numerous flexible self-sealing tanks located below the hull deck and in the center wing section. Each tank contained a fuel pump, fuel control valve and a quantity gauge. Using a single point refueling system, the *Marlin* could refuel at the rate of 400 gallons per minute at a twenty psi nozzle pressure. When not using a pressure system, the hull and auxiliary wing tanks could be gravity filled by conventional means. The *Marlin's* fuel system was further designed to jettison fuel at a rate of between fifty and seventy-five gallons per minute through a tube extending overboard at the hull stern post.

Designed as a flying boat, the *Marlin* required detachable beaching gear and ramp facilities for launch and retrieval from a land base. Using a three-point, individual gear system with attached floatation tanks, the beaching gear was capable of supporting a gross weight of 78,500 pounds with a 2-G safety factor. Attached on both sides of the aircraft at the waist position, the main beaching gear assemblies were color coded per side, green for starboard (right) and red for port (left). The tail beaching gear was secured at the aft end of the aircraft and was accessible from inside the tail compartment through a small hatch. Each main beaching gear was further equipped with a hand brake assembly for parking the aircraft.

Modification Program

Between 1957 and 1959, Martin was awarded several contracts to modernize, update and re-furbish the older P5M-1s and early P5M-2s in service with the Navy. Performed at the company's factory in Baltimore, the modification program affected nearly all *Marlins* in naval service, excluding current production models which included most updates in the manufacturing process.

Commencing in April 1957, the program consisted of structural updates, improvements in power plants and the installation of newer ASW electronics and specialized weapons systems. During the modification process, aircraft were completely disassembled, stripped, cleaned, repainted and flight tested before being returned to the Navy. By the time the aircraft were returned to their respective squadrons, they had been completely refurbished to "zero" time standards. Initially requiring 72,000 man-hours per aircraft, the modification process was refined and by 1959, the number had been reduced to nearly half the original figure.

Among the more obvious initial improvements was the installation of AN/APA-69A Direction Finding Equipment, identified by a *doghouse* enclosure atop the cockpit, and the AN/ASQ-8 Magnetic Anomaly Detector (MAD), mounted on top of the tall vertical stabilizer. Armament improvements included the installation of modified bomb carriers to accommodate aerial torpedoes and specialized internal weapons. Following the installation of additional ASW equipment, the basic crew complement was increased from seven to ten to include operators and support personnel.

The majority of the state-of-the-art ASW electronics modifications, retrofitted by Martin through 1959, were developed at the Naval Air Development Center, Johnsville, Pennsylvania. These updates were referred to as *Julie, Jezebel,* and Aircraft Integrated Display System (AIDS), and were designed primarily for the detection and localization of submerged submarines.

The *Julie*-ECM (Electronic Counter Measure), an explosive echo ranging and localization system, utilized a two-pen, paper trace ASA-20 recorder at the operator's station for furnishing range data. Data was fed into a computer (part of the AIDS) which computed the actual target position.

Jezebel, a search-detection system employed the use of *sonobuoys* (active or passive sonar sensors) for primary search detection and classification. Sonar data was transmitted to an ARR-58

Shown is a series of P5M-1s undergoing modification by Martin in 1958. (Martin Aircraft Company)

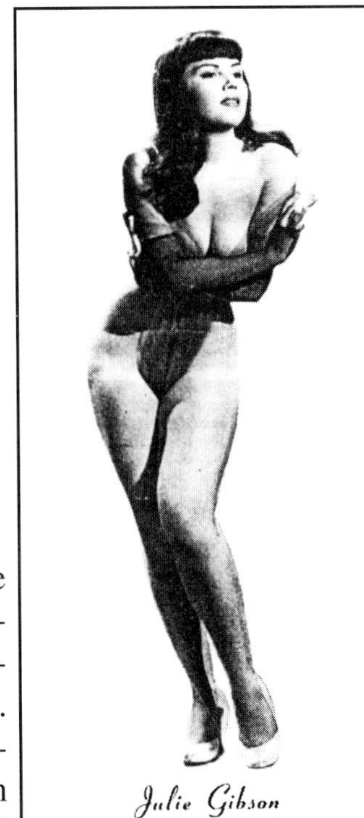

Pictured is *stripper* Julie Gibson, whose first name was given to the state-of-the-art electronic counter measures equipment. Many of the pieces of ASW equipment developed at NADC, Johnsville, Pennsylvania, were named for *strippers* in the Philadelphia area. (Courtesy of R. Greenwood)

Receiver Group and an AQA-3 Indicator Group located at the operator's position on the flight deck of the aircraft. The *Jezebel* system also had the capability to provide rough localization information.

Simply put, the AIDS was actually a data reduction center used to collect and compute data from other ASW equipment. Information was picked up, filtered and reduced by systems operators and sent to the Tactical Coordinator (TACCO), who combined the information in graphic form on the APA-125 indicator to determine aircraft-to-target range.

A new position in patrol aviation, designed around the ASW modifications, the Tactical Coordinator was typically a Non-Flight Officer (NFO) whose primary responsibility was to collect, evaluate, and correlate tactical intelligence information and direct tactical offensive or defensive measures to the pilot and crew. This information was presented to the pilot in graphic form on the pilot's ground track plotter. The introduction of the Tactical Coordinator position, along with the use of AIDS, were milestones in ASW tactics for that time and proved invaluable during operation *Market Time* off the coast of Vietnam between 1964 and 1967.

When coupled with the aircraft's MAD device, mounted high on either the forward edge of the vertical stabilizer on the P5M-1, or the trailing edge of the T-tail on the dash-two, the P5M possessed the latest technology in ASW capability. All aircraft having completed the modification program were assigned an "S" suffix letter to their primary designator, i.e., P5M-1(S) and P5M-2(S).

Production of the P5M-1 series continued until 1954 with a total of 167 produced in two basic configurations. The Navy accepted delivery of 160 ASW versions and seven were manufactured in the air-sea rescue configuration for the U.S. Coast Guard.

Aircraft were completely refurbished during the Modification Program, before being returned to the Navy. Shown is P5M-1S, BuNo 130277, upon delivery to FASRON-110 at NAS North Island. (Clay Jansson photo)

The Martin P5M
Marlin

Chapter Two

P5M-2

As the first of the P5M-1 models were being delivered to the fleet in 1952, Martin and Navy engineers were already working on a second generation of the *Marlin*. Designated the P5M-2, it offered significant design enhancements over the initial model. The most apparent changes were in the bow chine and tail assembly. The bow chine line was lowered to reduce spray height during landing and takeoff which helped to eliminate excessive spray from the propeller and flap areas. The tail assembly was completely redesigned to include a high-mounted T-tail with elevators set atop a tall, single vertical-fin stabilizer. This design proved to be the most efficient for both reduced weight and spray characteristics over the control surfaces and was a consideration on all future Martin flying boat concepts.

The design for this new model was not attributed to a single prototype version, but was the combined result of two different test aircraft. Once the P5M-1 model was accepted by the Navy, the XP5M-1 prototype was designated the M-270 and used for research and development of a new concept in hull design for future models. These tests proved the lower bow chine theory, and it was incorporated in the P5M-2 model. However, the development of the T-tail was not part of the M-270 project. Confident in the initial design, Martin modified a single production P5M-1 to include the new T-tail assembly. In August 1953, Navy BuNo 130286, a standard production model P5M-1, but with the new T-tail assembly added, took to the air for initial flight tests. Having been proven successful, the new design was used on all production models of the P5M-2.

The first official flight of a production P5M-2 took place on April 29, 1954, with the Navy accepting delivery of their first aircraft (BuNo 135475) on June 23, 1954. This particular aircraft was the second production P5M-2 and the oldest to see continuous active service with the Navy. It continued its service life with various squadrons throughout the Vietnam conflict until being scrapped in Japan, in June 1967. Fleet delivery of the new model commenced on October 18, 1954, with VP-47 based at NAS Alameda as the first squadron to receive them.

The engines used on the P5M-2 were uprated to the improved Curtiss-Wright Turbo-Compound R-3350-32WA, producing 3,450 horsepower each and giving the dash-two a slight boost in maximum performance to 276 miles per hour (240 knots) indicated airspeed. Range remained similar to the earlier model at approximately 3,000 miles with an increased normal gross weight of 78,000 pounds.

Shown is a P5M-2, BuNo 135477, during flight acceptance testing in Baltimore. The second generation *Marlin* offered significant improvements over the earlier P5M. (U.S. Navy photo)

This photo shows the lower bow chine used on the P5M-2. This was the most significant improvement in the initial hull design of the P5M. Distance from the ground to propeller tip is over 12 feet. Propeller diameter is 15 feet. (U.S. Navy photo)

This June 1953 photo shows P5M-1, BuNo 130286, undergoing modification of the prototype T-tail concept used on the P5M-2. This aircraft served as the only prototype for the design. (U.S. Navy photo)

On display at the National Museum of Naval Aviation, Pensacola, Florida, is this wind tunnel model of a P5M-1 with the dash-two tail modification. (Author's Collection)

The Navy accepted delivery of the first P5M-2, BuNo 135475, in ceremony at NAS Norfolk on June 23, 1954. This particular aircraft flew continuously with the Navy until being scrapped in 1967, following service in Vietnam. (U.S. Navy photo)

This 1954 photo of BuNo 135477 shows the cavernous bomb bays, tail turret and MAD boom installation on the P5M-2. (U.S. Navy photo)

Shown is the third production P5M-2, BuNo 135476, outside a Martin hanger at Strawberry Point in June 1954.
(U.S. Navy photo)

This October 1954 photo shows BuNo 135477 making a perfect landing, following flight testing in the Baltimore area.
(U.S. Navy photo)

Stores loading capacity increased proportionally to 10,000 pounds.

Crew accommodations in the second generation *Marlin* were again of primary importance and were refined from those of the earlier model. More comfortable, adjustable seating was added, as were additional berthing facilities which were moved aft and out of the immediate vicinity of the noisy APU, providing a quieter environment for resting personnel. The crew complement on the dash-two was increased to thirteen, in order to accommodate operators for the new ASW electronic equipment and additional technical support personnel.

Although the dash-two version had somewhat larger overall dimensions than that of the earlier dash-one, the ten foot width remained the same, as did the number of internal compartments. The interior was divided into fourteen compartments; bow compartment, forward entrance compartment, electronics compartment, main beaching gear compartment, pilot's compartment, flight deck compartment, fuel compartment, APU compartment, sonobuoy compartment, forward waist compartment, aft waist compartment, stowage compartment, hydraulics compartment and tail gunner/observer's compartment.

The lower internal compartments were separated by three watertight doors to prevent centralized flooding in the event of structural damage to the aluminum hull. These were installed in the bulk-

This 1955 view of the P5M-2, BuNo 135475, shows the clean lines of the T-tail modification and the 70,000,000 candle power search light installed on the right wing tip. (U.S. Navy photo)

ANTENNAS

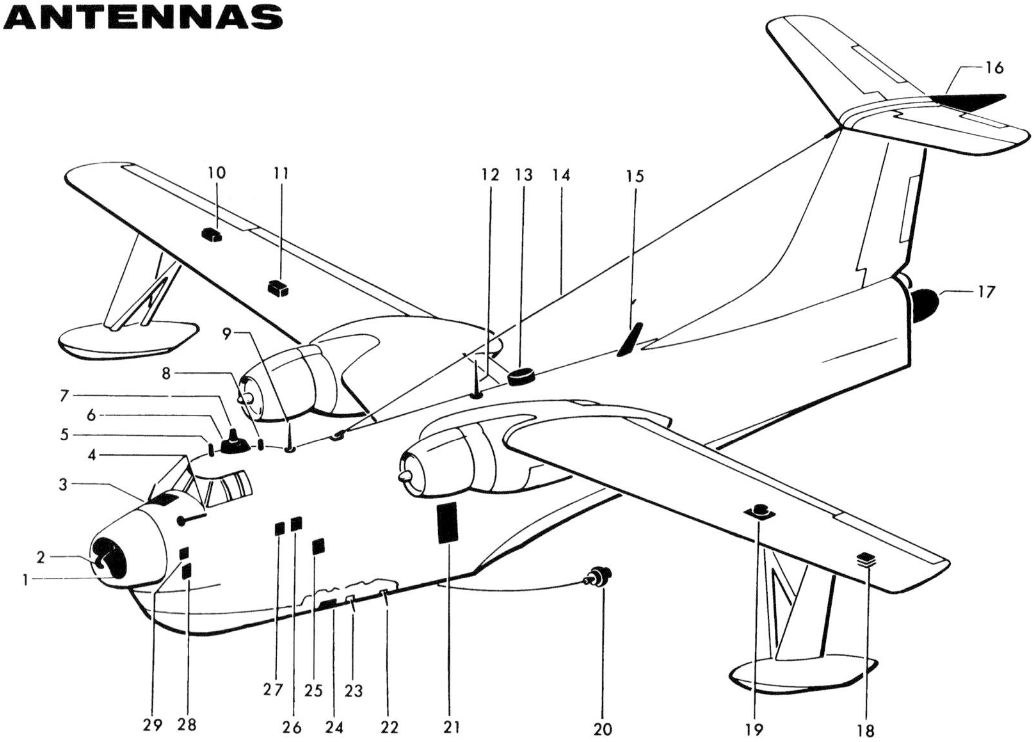

ANTENNA	ASSOCIATED SYSTEM
1. AS–1000/APS–80	AN/APS–80
2. AN/APX–7 (integral part of AS–1000/APS–80)	AN/APX–7
3. AS–578/ARA–25	AN/ARC–27A AN/ARR–26 (receiver B)
4. AT–563/AP	AN/ALR–8
5. AT–141A/ARC	AN/ARC–27A
6. AS–676/APA–69	AN/APA–69C
7. AS–133/APX–6	AN/APX–6
8. AT–500/AP	AN/ALR–8
9. AT–8/AR	AN/ARC–1
10. R–88–T–1950–50 (compass transmitter)	G2 COMPASS
11. RT–160/APN–22	AN/APN–22
12. AT–563/ARN	AN/ARN–6
13. AS–313/ARN	AN/ARN–6
14. FIXED WIRE ANTENNA	AN/APN–4 AN/ARC–38 AN/ARR–41
15. AT–121/AP	AN/ALR–8
16. DT–37/ASQ–8	AN/ASQ–8
17. RT–533/APN–122 (receiver-transmitter)	AN/APN–122

ANTENNA	ASSOCIATED SYSTEM
18. AT–134A/ARN	AN/ARN–12
19. KJ–4 (compass transmitter)	MF–1 COMPASS
20. TRAILING WIRE ANTENNA	AN/APN–4 AN/ARC–38 AN/ARR–41
21. AT–495/ARN–14 (port and starboard)	AN/ARN–14E
22. AT–466/ARR	AN/ARR–26 AN/ARR–58 (receiver A and C)
23. AT–466/ARR	AN/ARR–58 (receiver B and D)
24. E–845012	AN/ARN–21
25. AT–302/APR–9 (port and starboard)	AN/ALR–8
26. AT–303/APR–9 (port and starboard)	AN/ALR–8
27. AT–304/APR–9 (port and starboard)	AN/ALR–8
28. AT–303/APR–9 (port and starboard)	AN/ALR–3
29. AT–304/APR–9 (port and starboard)	AN/ALR–3

TABLE CXVII–A
STORES LOADING

—EXTERNAL STORES—

QUANTITY	STORES	ACTUAL WEIGHT
2	Mark 24 Mine (700 lb)	700
4	Mark 19–2 Mine (500 lb)	540
2	Mark 36–1, –2, –3 Mine (1000 lb)	1065
4	Mark 54–1 Depth Bomb	355
4	Mark 15–3 Practice Bomb (100 lb)	100
4	Mark 65–0 Practice Bomb (500 lb)	388
2	Mark 66–0 Practice Bomb (1000 lb)	859
4	AN–M30A–1 G. P. Bomb (100 lb)	110
4	AN–M57A–1 G. P. Bomb (250 lb)	252
4	AN–M–64A–1 G. P. Bomb (500 lb)	545
2	AN–M–65A–1 G. P. Bomb (1000 lb)	994
4	AN–M–58–A–2 SAP Bomb (500 lb)	530
2	AN–M–59–A–1 SAP Bomb (1000 lb)	1024
4	AN–M–81 Fragmentation Bomb (260 lb)	262
4	AN–M–88 Fragmentation Bomb (220 lb)	223
4	Mark 3–0 Smoke Bomb (100 lb)	102
4	AN–M–78 Chemical Bomb (500 lb)	490
2	AN–M–79 Chemical Bomb (1000 lb)	937
4	AN–47A–3 Incendiary Bomb (100 lb)	72
4	AN–M–14 Aimable Cluster (500 lb)	463
4	AN–M–17A–1 Aimable Cluster (500 lb)	463
4	AN–M–76 Incendiary Bomb (500 lb)	475
8	5-inch HVAR	140
8	Aero 6B Rocket Launcher Package	150

—INTERNAL STORES—

QUANTITY	STORES	ACTUAL WEIGHT
4	Mark 24 Mine (700 lb)	700
4	Mark 19–2 Mine (500 lb)	540
4	Mark 36–1, –2, –3 Mine (1000 lb)	1065
4	Mark 54–1 Depth Bomb	355
4	Mark 15–3 Practice Bomb (100 lb)	100
4	Mark 65–0 Practice Bomb (500 lb)	388
4	Mark 66–0 Practice Bomb (1000 lb)	859
4	AN–M–30A–1 G. P. Bomb (100 lb)	110
4	AN–M–57A–1 G. P. Bomb (250 lb)	252
4	AN–M–64A–1 G. P. Bomb (500 lb)	545
4	AN–M–65A–1 G. P. Bomb (1000 lb)	994
4	AN–M–58A–2 SAP Bomb (500 lb)	530
4	AN–M–59A–1 SAP Bomb (1000 lb)	1024
4	Mark 33–0–AP Bomb (1000 lb)	1012
4	AN–M–81 Fragmentation Bomb (260 lb)	262
4	AN–M–88 Fragmentation Bomb (220 lb)	223
4	Mark 3–0 Smoke Bomb (100 lb)	102
4	AN–M–78 Chemical Bomb (500 lb)	490
4	AN–M–79 Chemical Bomb (1000 lb)	937
4	AN–47A–3 Incendiary Bomb (100 lb)	72
4	AN–M–14 Aimable Cluster (500 lb)	463
4	AN–M–17A Aimable Cluster (500 lb)	463
4	AN–M–76 Incendiary Bomb (500 lb)	475
4	Mark 34–1 Torpedo (1070 lb)	1157
4	Mark 10–9 Mine (1860 lb)	1860
4	Mark 25–1 Mine (2000 lb)	1930
4	Aero 4–A Bomb Container	75
4	Mark 47–3 Practice Bomb Packard Container	62
4	Aero 5A Practice Bomb Container	
4	AN–MK–1–0 AP Bomb (1600 lb)	1585
4	AN–M–5 Fragmentation Cluster (100 lb)	87
4	AN–M–12 Incendiary Cluster (100 lb)	105
4	AN–M–6 Incendiary Cluster (100 lb)	140
4	AN–M–13 Incendiary Cluster (500 lb)	427
1	Mark 90 Depth Bomb	2497
4	Mark 43–1 Torpedo	265
2	Mark 101 Depth Bomb	1185

MAXIMUM STORES LOADING FOR EACH SIDE OF AIRPLANE

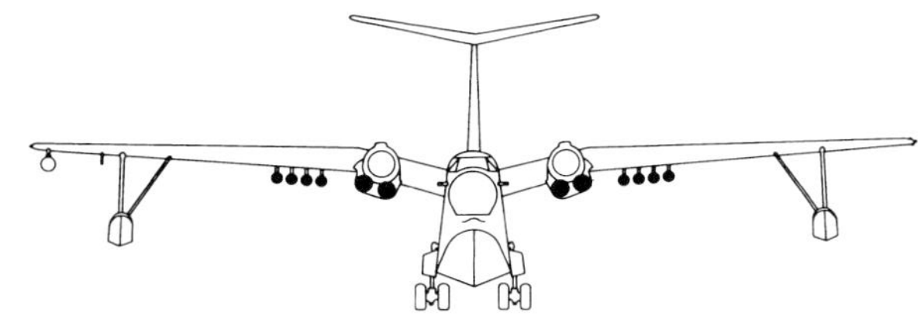

	QUANTITY	UNIT WEIGHT	STORES
● (small)	4	262 LB.	EXTERNAL
● (large)	2	2050 LB.	INTERNAL

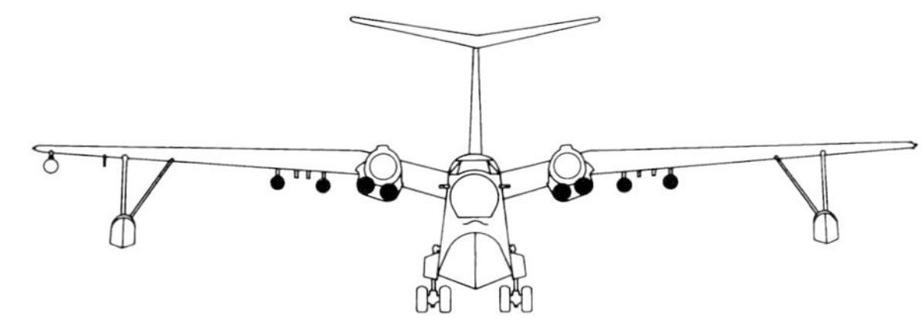

	QUANTITY	UNIT WEIGHT	STORES
● (small)	2	545 LB.	EXTERNAL
● (large)	2	2050 LB.	INTERNAL

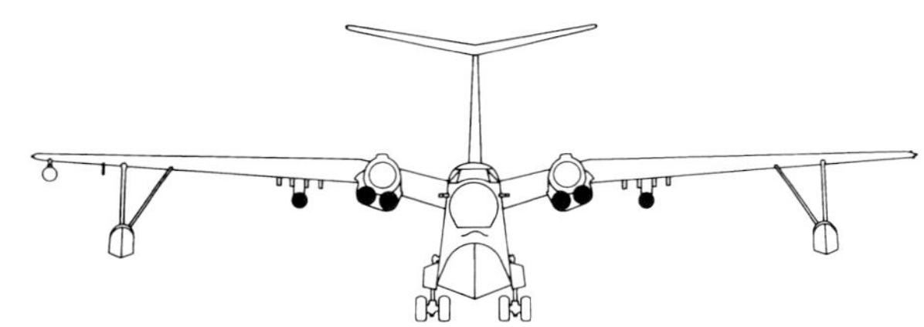

	QUANTITY	UNIT WEIGHT	STORES
● (small)	1	1065 LB.	EXTERNAL
● (large)	2	2050 LB.	INTERNAL

MAXIMUM LOAD ON EXTERNAL CARRIERS SHALL NOT EXCEED 1090 LBS.
MAXIMUM LATERAL AIRPLANE UNBALANCE SHALL NOT EXCEED 128,800 FT. LBS.

PILOTS' COMPARTMENT

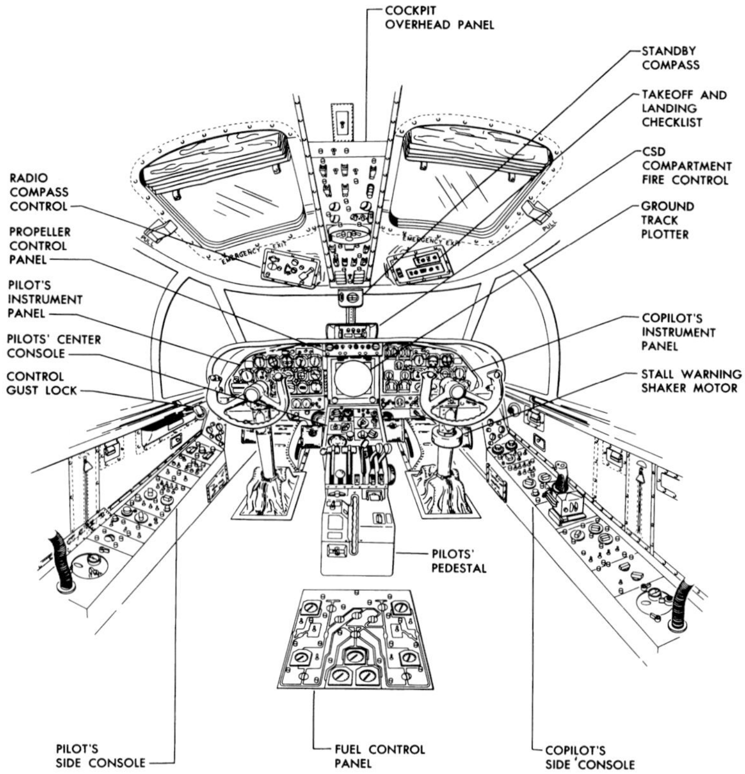

FUEL STOWAGE

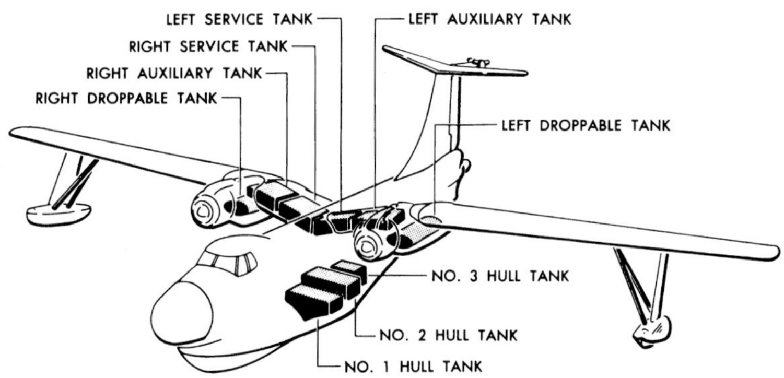

BEFORE P5M–ASC–448		FUEL TANK	AFTER P5M–ASC–448	
GALLONS	POUNDS		GALLONS	POUNDS
262	1572	LEFT SERVICE	262	1572
265	1590	RIGHT SERVICE	265	1590
510	3060	LEFT AUXILIARY	510	3060
508	3048	RIGHT AUXILIARY	508	3048
575	3450	LEFT DROPPABLE	575	3450
575	3450	RIGHT DROPPABLE	575	3450
285	1710	NO. 1 HULL	295	1770
571	3426	NO. 2 HULL	584	3504
408	2448	NO. 3 HULL	419	2514
3959 GALS	23,754 LBS		3993 GALS	23,958 LBS

GENERAL ARRANGEMENT

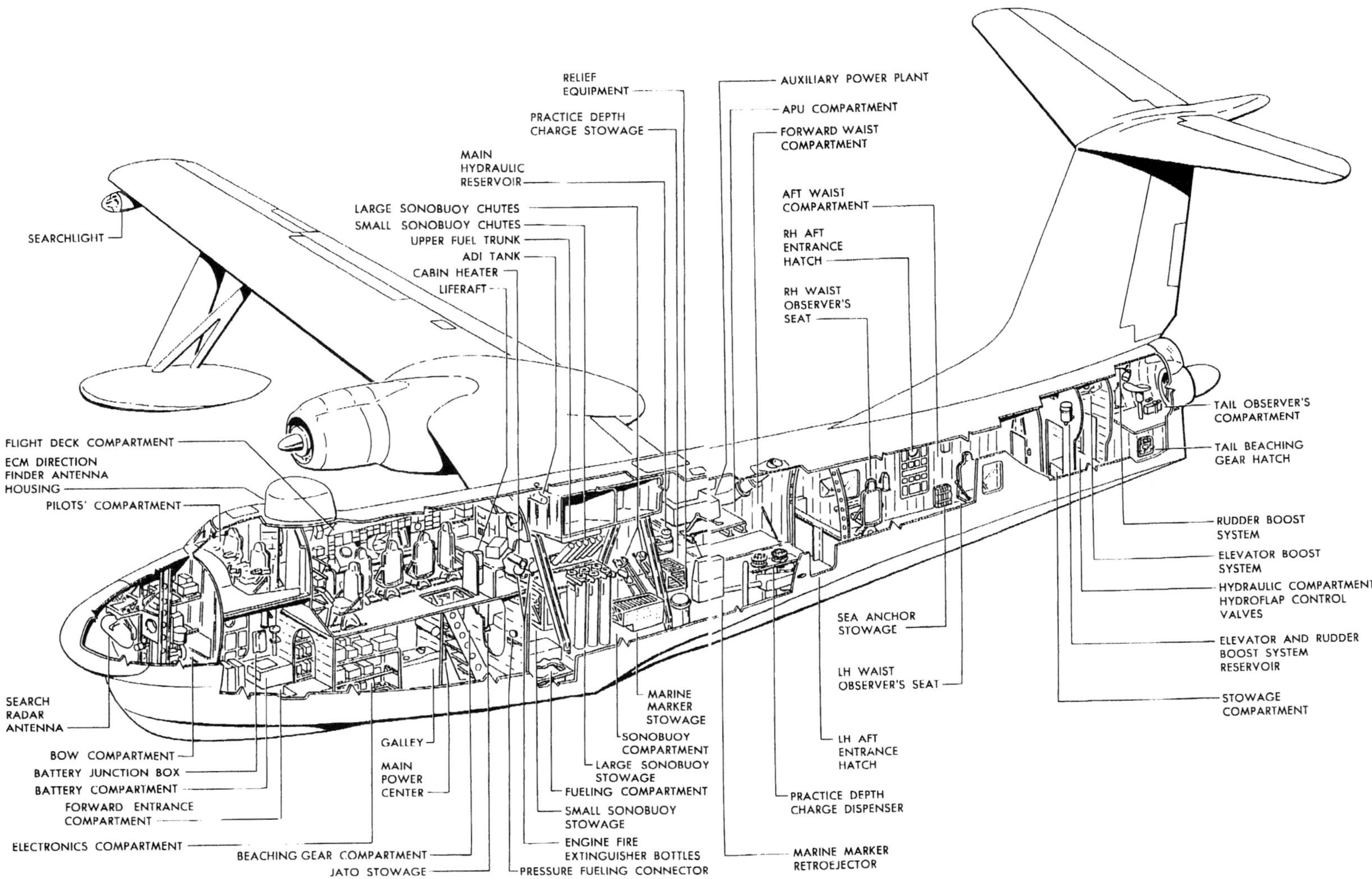

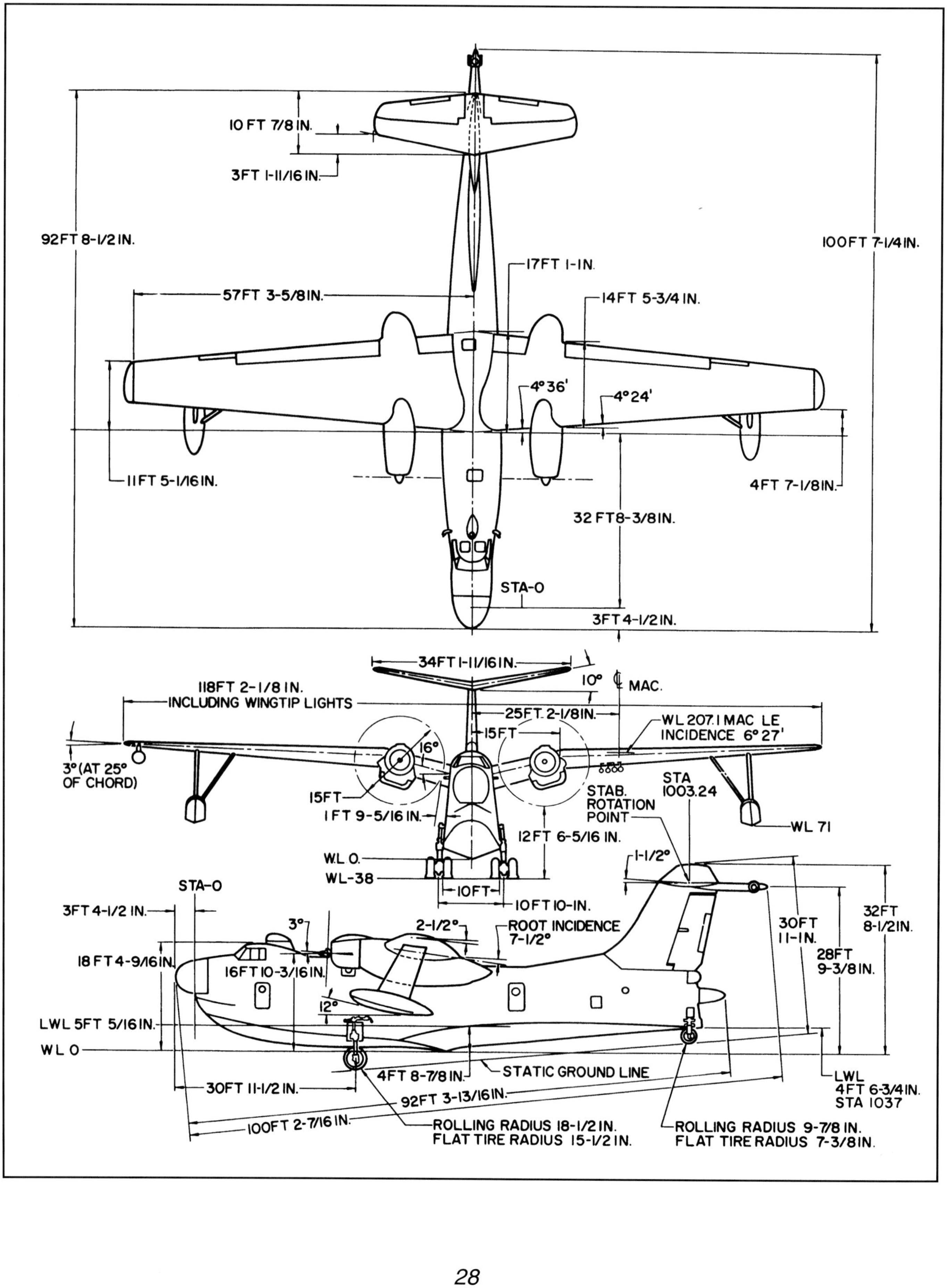
10 FT 7/8 IN.
3FT 1-11/16 IN.
92FT 8-1/2 IN.
100FT 7-1/4 IN.
17FT 1-IN.
57FT 3-5/8 IN.
14FT 5-3/4 IN.
4°36'
4°24'
11FT 5-1/16 IN.
4FT 7-1/8 IN.
32 FT 8-3/8 IN.
STA-0
3FT 4-1/2 IN.
34FT 1-11/16 IN.
10°
℄ MAC.
118FT 2-1/8 IN.
INCLUDING WINGTIP LIGHTS
25FT. 2-1/8 IN.
WL 207.1 MAC LE
INCIDENCE 6° 27'
15FT
16°
3°(AT 25° OF CHORD)
15FT
1 FT 9-5/16 IN.
STAB. ROTATION POINT
STA 1003.24
WL 71
12FT 6-5/16 IN.
WL 0.
WL-38
10FT
10FT 10-IN.
1-1/2°
STA-0
3FT 4-1/2 IN.
3°
2-1/2°
ROOT INCIDENCE 7-1/2°
18 FT 4-9/16 IN.
16FT 10-3/16 IN.
12°
30FT 11-IN.
32FT 8-1/2 IN.
28FT 9-3/8 IN.
LWL 5FT 5/16 IN.
WL 0
4FT 8-7/8 IN.
STATIC GROUND LINE
30FT 11-1/2 IN.
92FT 3-13/16 IN.
100FT 2-7/16 IN.
LWL 4FT 6-3/4 IN. STA 1037
ROLLING RADIUS 18-1/2 IN.
FLAT TIRE RADIUS 15-1/2 IN.
ROLLING RADIUS 9-7/8 IN.
FLAT TIRE RADIUS 7-3/8 IN.

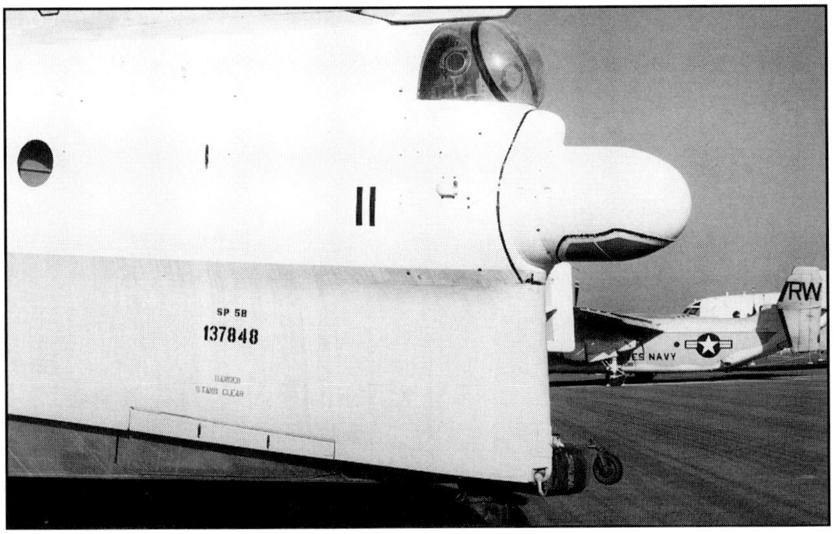

This rear view of an SP-5B, BuNo 137848, shows the antenna for the AN/APN-122 system which replaced the twenty-millimeter tail turret. The gunner's window was retained as an observer's position.
(Clay Jansson photo)

Modified to accommodate oversize ordnance, the left bomb bay on the SP-5B was fitted with deeper doors, as shown here on BuNo 137848 at NAS North Island in 1967.
(Clay Jansson photo)

This view of the *Marlin's* T-tail shows the installation of the Magnetic Anomaly Detector (MAD) boom which sat atop the tail away from magnetic material.
(Clay Jansson photo)

head at the aft end of the forward entrance compartment, at the aft end of the electronics compartment, and at the aft end of the beaching gear compartment.

The electronics suite of the P5M-2 included an improved, more powerful radar system, the AN/APS-80. The largest airborne search radar of its kind at the time, it was capable of detecting an object as small as a submarine periscope at a considerable distance. Other standard improvements in ASW electronics included the latest technology in counter measures and detection equipment, the same type equipment as was retrofitted on earlier models. Although the P5M-2 initially retained the defensive remote controlled twenty-millimeter turret in the tail, it was later deleted on most all models by 1958 in favor of additional ASW detection equipment.

Under the modification and update program, performed between 1957 and 1959, most dash-two models were modified to accommodate special weapons, including the Mark 90 and Mark 101 nuclear depth bomb. Designed for use only in the left bomb bay, the modifications included revised bomb carriers and deeper bomb bay doors to facilitate the large depth bombs, weighing 2,500 and 1,200 pounds, respectively.

Shown at NAS North Island in 1966 is an SP-5B, BuNo 141258. It was modified with a small jet-engine in the tail in an attempt to extend the operational life of the *Marlin.* The project was not accepted by the Navy and was terminated in 1966.
(Clay Jansson photo)

Series Conclusion

Production of the P5M-2 continued through December 1960, producing a total of 116 aircraft in two base model configurations. The Navy received 102 advanced ASW versions and the U.S. Coast Guard took delivery of four P5M-2G models equipped for air-sea rescue. In 1959, ten aircraft were configured with basic ASW equipment for use by the French under a lend-lease program with the United States. Incidentally, France was the only nation, other than the U.S., to operate the *Marlin.*

Among those designs considered during the peak of seaplane development in the late 1950s were projects that included the use of jet engines, turbo-props, boundary layer control, dipping sonar, STOL capabilities and improved hull designs to allow for operation in higher sea-states. Although most of these designs proved both feasible and functional, none found their way into full production or were incorporated in the existing P5M design. Several nations including France, New Zealand and Japan had initially submitted orders for the P5M-2, however none actually followed through with formal delivery commitments.

In 1964, an attempt was made to extend the useful service life of the *Marlin* by mounting a small jet-engine in the tail to improve performance. Installed and evaluated on a single SP-5B, the modification came too late to prolong the life of the naval seaplane and the project was abandoned in 1966.

Although Martin continued to explore new technology in seaplane design, the SP-5B was the last model accepted and produced for the Navy. It was also the last flying boat manufactured and produced by the Martin Aircraft Company. In nearly sixteen years of operation, the Martin P5M series proved itself a venerable replacement for the vintage PBM of World War Two. Over the years, several new designs were considered to replace the P5M, and in some cases prototypes were built, such as the jet-powered P6M *Seamaster.* But, for numerous reasons that seemed politically motivated and not technology driven, production of naval seaplanes ceased abruptly in 1960 and then vanished from the Navy's inventory in 1967.

This P6M-1 *Seamaster,* BuNo 143823, shown in 1959, is one of less than twelve built in the P6M series before the project was canceled later in 1959. None have survived. (U.S. Navy photo)

The Martin P5M
Marlin

Chapter Three

Martin P5M-3 Proposal

Advancement in naval seaplane design continued well into the late 1950s with several manufacturers participating in competition for an advanced design to replace the P5M-2. As the predominant manufacturer of seaplanes for the Navy, the Martin Aircraft Company already had two models in production (*Mars* and *Marlin*) and were currently developing the P6M *Seamaster.* The four-engine, jet-powered, *Seamaster* was the first seaplane design capable of speeds in excess of 600 miles per hour. In addition to their production and development efforts, Martin submitted a proposal to replace the P5M-2 in 1955. Designated the P5M-3, this conventionally powered, four-engine entry offered significant improvements in hull design and performance over the earlier twin-engine dash-two model *Marlin.*

The concept of a conventionally powered, four-engine, medium-range flying boat was not new to Martin. In 1945, Martin proposed an advanced, four-engine model with a similar long after-body hull design to that of the XP5M-1. Designated Model 230-6, the proposal was not accepted by the Navy and the project was canceled.

Designed as an advanced open-ocean ASW aircraft with STOL (Short Take Off and Landing) capabilities, the proposed P5M-3 was powered by four, Wright R-1820-82 radial engines and incorporated the latest in ASW electronics, improved beaching techniques and exceptional crew accommodations. Using the M-270 and P6M and based on extensive engineering and research in hull design, the P5M-3 was proposed to be capable of repeated takeoff and landings under sea-state five (eight to twelve foot waves) conditions.

One of the more unique features of the P5M-3, was the introduction of Boundary Layer Control (BLC) to improve open sea performance. Powered by a single GE-XJ85 jet-engine mounted top-center of the wing, air from the 2,400-pound thrust turbine flowed through a series of ducts that exhausted through a continuous span of nozzles over the top of the flaps. Along the entire wing trailing edge, boundary air was in effect smoothed out, thereby reducing stall speed to between forty-three and forty-eight miles per hour, respectively, for takeoff and landing at gross weight. Used during takeoff, BLC could effectively reduce the required distance by nearly fifty percent. The use of *hydroflaps* was deleted on the P5M-3 and replaced by ducted air from the BLC system on each side of the lower aft fuselage. In this capacity, BLC provided the same waterborne maneuverability as did conventional *hydroflaps* pioneered by Martin on earlier models.

Basic crew complement for the proposed model consisted of six: pilot, co-pilot, navigator-tactical coordinator, two ASW operators and a plane captain. Accommodations included improved seating, capable of withstanding higher accelerations,

This 1945 concept drawing of the Martin Model 230-6 was not accepted by the Navy as a replacement for the PBM. (Martin Aircraft Company)

Under the designation P5M-3, this new design for a four-engine medium-range flying boat was submitted in 1955. (Martin Aircraft Company)

Shown under construction in November 1956 is the full-scale mock-up of the proposed P7M *SubMaster*. (Martin Aircraft Company)

berthing for four and a well equipped galley area. Overall cockpit visibility was improved through use of a new wrap-around windshield and large convex windows in the flight deck area.

Beaching of the P5M-3 was proposed by using a *beaching vehicle* or cradle (similar in concept to that used by the P6M *Seamaster*), whereby the aircraft taxied directly into the vehicle and up the ramp under its own power. An alternate design allowed for a retractable, tricycle beaching gear system that could be used on land, snow or ice.

Based on a favorable review of the proposal, submitted in December 1955, Martin started work on a full-scale mock-up of the new four-engine concept. Initially dubbed the P5M-3 in early 1955, the project was later assigned the name *SubMaster* and designated the P7M, to conform to model number sequence. The aluminum and wood engineering mock-up was a full-scale representation of the new design. From its finely detailed interior layout to its sleek exterior, painted in Navy trim, the new concept looked promising. Although completed on schedule in late 1956, it was not accepted by the Navy and no prototype was built. Other competing companies, including Grumman and Convair, submitted similar proposals during this period. The winning design, designated the P6Y, was awarded to Convair, however the Navy canceled the contract before a prototype could be completed. The cancellation of this contract and subsequent termination of the P6M project in August 1959, signaled the end of future development of naval seaplanes.

Martin P5M-3
Proposed Performance Summary

Gross Weight:		Engines:	
Takeoff:	67,400 pounds	Main: 4 each	Wright R-1820-82
Combat:	59,316 pounds	BLC: 1 each	GE-XJ85
Landing:	62,028 pounds	Speed: Maximum:	250 mph
		Cruise	147 mph
Maximum Fuel:	3,585 gallons	Stall: Landing/gross weight:	
Normal Payload:	2,367 pounds	with BLC	48 mph
Armament:	6 x Mk 43-1 torpedoes	without BLC	62 mph
	8 x 2.75 in. rocket launchers	Rate of Climb:	1,300 feet/minute
Takeoff Wing Loading:	36.4 pounds/square foot	Range:	2,500 nautical miles
Service Ceiling:	20,700 feet	Length:	110 feet 5 inches
Wingspan:	136 feet	Height:	37 feet 4 inches
Takeoff Time: w/ BLC	12 seconds	Distance: w/ BLC	530 feet
without BLC	19 seconds	without BLC	1,000 feet

Martin P5M-3
Proposed ASW Electronics

System / Equipment Type	Designation	Purpose
Radar:	AN/APS-42	Surface Detention
Electronic Counter Measure:	AN/ALD-2	Instantaneous Direction Finder
	AN/ALA-3	Pulse Analyzer. Detect & Classify Enemy Radar
Magnetic Anomaly Detector:	AN/ASQ-10 (XN-2)	Localization Equipment
Sonar:	Developmental -	Active & Passive Dipping Sonar
Range: 25 miles	AN/AQS-6 (XN-2)	using single 75 inch by 80 inch
Depth: zero to 250 feet		transducer array
Sonobuoys:	AN/SSQ-2, -15, -20	Detection

Shown during flight testing in October 1967 is the first production prototype PS-1, s/n 5801 which combined seaplane technology from several nations and was considered the most advanced design of its time.
(Shin Meiwa Industries)

The second production prototype PS-1, s/n 5802, is shown during sea trials in June 1968.
(Shin Meiwa Industries)

Shin Meiwa Industries, Ltd.

During the Second World War, the Kawanishi Aircraft Company produced such famous flying boat designs as the Kawanishi H8K2 *Emily* and *Mavis,* as they were known by allied forces. In 1949, the company was re-established as Shin Meiwa Industries, Ltd. With factories in Itami and Konan, Shin Meiwa employed over 2,000 and was engaged primarily in repair and overhaul of U.S. military and commercial aircraft. They were awarded a contract in the early 1950s to research and develop a flying boat design for the newly established Japanese Maritime Self Defense Force (JMSDF). By 1966, they were in the final development stages of completing their first prototype, four-engine turboprop multi-role amphibian, the PX-S. This premier design employed the latest in flying boat technology developed by several nations, including the United States, Great Britain and Canada and had the distinction of being the world's most advanced seaplane.

Contrary to earlier writings, the P5M itself played no significant part in the prototype development of this aircraft and had only an indirect influence on its design. The actual prototype for the design was developed using a single, highly modified, Grumman UF-1 *Albatross* amphibian (BuNo 149822). Other than a *distinctly obvious* resemblance to a Martin design, and sharing some common components with the P5M by application, the PS-1, as it was to be called, borrowed no structural or other components from the salvaged Navy P5Ms at the facility.

The highly modified ex-JMSDF *Albatross,* designated the UF-XS, was fitted with an early version of the Spray Suppresser and BLC system. Further modified from a two-engine to a four-engine configuration, the three-quarter scale model prototype made its first flight in December 1962. The addition of two Pratt & Whitney R-1340-AN-1 engines provided additional conventional power and the installation of a General Electric T58-GE-6 turbine produced the blow-down air necessary for the BLC system.

Assigned the final designation, PS-1, construction of the first production prototype commenced in 1965 and was completed in the fall of 1967. The full-scale prototype PS-1 (S/N 5801) departed Kobe harbor on its maiden flight on October 29, 1967. Production of the second prototype (5802) commenced in 1966 and was completed in May 1968 with its first flight on June 14, 1968.

Designed as a flying boat, the PS-1 was uniquely fitted with an internal beaching gear system, enabling it to be launched and beached under its own power. Powered by four 2,850 shaft horsepower General Electric T64-IHI-10 turboprops, driving Hamilton Standard 63E60-15 three-blade constant-speed reversible pitch propeller's, the PS-1 incorporated two additional unique features in its STOL design.

• Spray Suppresser: A filleted, slatted, aerodynamic groove built around the bow of the seaplane's low-shock hull. During takeoff and landing, water is channeled through the groove and expelled at a point aft of the wings trailing edge, well behind the area which would ultimately effect the propellers and engines.

• Boundary Layer Control: The BLC system was powered by a single, General Electric T58-10-M1 three-stage, axial-flow compressor with 1,400 shaft horsepower. Installed in the upper center fuselage section, just aft of the wing, this power plant provided blow-down air which when distributed over the control surfaces of the wing, horizontal stabilizer and rudder during takeoff and landing, could in effect triple the lift of the wing. With BLC in operation, the PS-1 could take-off at speeds as slow as fifty-two miles per hour (45 knots) at a gross weight of 75,000 pounds. Further studies with the BLC showed the PS-1 capable of operating in higher sea-states of up to fourteen foot waves with thirty mile per hour surface wind conditions.

In service with the JMSDF, the multi-role seaplane was produced in two standard models, the PS-1 and the US-1. The PS-1 is configured for ASW use and employs a different retractable gear system than that of the US-1. The PS-1 is fitted with an internal retractable beaching gear system that allows for launch and retrieval of the aircraft under its own power. The amphibious, US-1 version is equipped with a retractable landing gear system that can be used on conventional runways. Configured as an air-sea rescue aircraft, the US-1 is painted the same Insignia White over Light Gull Gray as that of the PS-1, but with international orange markings (unarmed) on the nose and tail assembly.

A single PS-1 model, serial number 5801, was later converted for use as a fire-bomber for Japan's National Fire Protection Service.

Shin Meiwa PS-1 Specifications

Wing Span:	110 feet
Length:	110 feet
Height:	33 feet
Weight, Empty:	52,700 pounds
Normal Gross Weight:	82,500 pounds
Maximum Gross Weight:	99,000 pounds
Maximum Speed:	345 mph
Cruise Speed:	160 - 288 mph
Range:	2,800 miles
Main Engine: 4 each	GE T64-IHI-10
BLC Engine:	GE T58-10M1
Maximum Fuel:	6,350 gallons
Crew:	8 to 12

Shown at NAF Atsugi, Japan in 1986, this US-1 belongs to Flight Squadron 71, *Rescue Ivory.* Note the international orange markings on the nose and tail which are similar to those used on U.S. *Marlins* for training. (Edward J. Miles photo)

The Martin P5M
Marlin

Chapter Four

Research and Development

Built specifically for ASW applications, the Martin P5M was assigned primarily to operational fleet squadrons throughout its useful service life. As the Navy's only post-war production medium-range seaplane however, the P5M was also used for various hydrodynamic projects and product evaluations relative to future development of naval seaplanes.

During the 1950s, the P5M was used extensively in the development and testing of future ASW weapons, systems, seaplane designs, and advanced-base concepts to extend the capabilities of the naval seaplane. Squadrons routinely participated in developmental programs and exercises, involving advanced water-based docking systems, submarine refueling, and surface support ships. While pioneering new concepts for the future, these tests offered a learning experience to determine what could and could not be done with seaplanes. While the application of the naval seaplane remained the same, versatility, performance, and hull design became the most critical factors in the development of future waterborne aircraft.

Typically, new models required a prototype; however due to the success and strength of the P5M design concept, Martin was able to re-configure the XP5M-1 prototype for continuing engineering programs. Used primarily as a development tool for future designs, it virtually eliminated the need for costly interim prototypes.

Working with Martin, the Navy continued to use the Model 237 concept on engineering development programs, effecting both future aircraft design and seaplane support projects. Throughout the 1950s and early 1960s, several research and development projects were undertaken using the M-270, P5M-1 and -2. The majority of these projects were developed and engineered jointly between the Martin Aircraft Company; Stevens Institute of Technology, Hoboken, New Jersey; Air Development Squadron One (VX-1), Key West, Florida; and the Naval Air Test Center (NATC), Patuxent River, Maryland.

Shown here are the two primary aircraft used by Martin for research and development of seaplane technology during the 1950s. Both the M-270 and a single Grumman J4F-2, nicknamed the *Petulant Porpoise,* were modified with a special 15 to 1 hull design and played equally important roles in future seaplane development. (U.S. Navy photo)

Test and Development Aircraft

✪ Martin M-270: BuNo 98616

Following the Navy's acceptance of the P5M-1, the XP5M-1 prototype was further modified and used as a test-bed for future seaplane development. Redesignated research model M-270 in late 1950, the highly modified PBM-5, employing a proven wing design and engines, was fitted with a new style hull, adding nearly one foot to its existing length. Based on extensive hull research at the Stevens Institute, modifying the existing XP5M reduced the need to build a completely new prototype at a significant cost savings. Foreseeing a need for continued engineering in longer, slimmer hull designs for superior aerodynamic seaplane shape and more efficient structural bodies, while retaining or improving on existing designs, flying boat engineers at Martin and Stevens Institute worked towards building a full-scale model for water and flight testing.

Incorporating an optimum configuration for hydrodynamic, aerodynamic and structural weight, with a length-to-beam ratio of 15 to 1, the extensive modification to the hull of the M-270 began in March 1951. Completion of the retrofit and subsequent flight testing resulted in revised hull designs that were used on the P5M-2 and P6M *Seamaster*.

The prototype hull design used on the M-270 was also mated to a specially modified Grumman J4F-2 *Widgeon*. Nicknamed the *Petulant Porpoise*, this aircraft played an equally important role in the development and evaluation of future hull designs for naval flying boats.

Shown at Strawberry Point in May 1952 is the highly modified XP5M-1 which was redesignated Model 270 in early 1951. As the M-270, it was converted to evaluate advanced hull designs for the P5M-2 and P6M. (U.S. Navy photo)

Although extensively modified, the Model 270 retained the original PBM wing and floats designed in 1937. (U.S. Navy photo)

The aging XP5M, which started life as a standard production PBM-5 in 1946, was Martin's primary test aircraft during the post-war development of medium-range military seaplanes and advanced hull designs. The M-270 modification was its last project.

✪ P5M-1: BuNo 130286

Modified in June 1953 with a high-mounted T-tail, this standard production P5M-1 served as the primary prototype test aircraft in the development of a new tail concept used on the P5M-2. Following evaluation and the Navy's acceptance of the new tail design, the aircraft was returned to the fleet in its modified state. Still classified as a P5M-1, it continued its operational service life with at least two known patrol squadrons, VP-46 and VP-48, before being removed from naval service at NAS North Island in early 1967.

✪ P5M-2: BuNo 140146 and 140147

Manufactured in 1957, these two aircraft were used by VX-1 in the evaluation of at least two known developmental projects, between 1957 and 1958. (Note: VX squadrons were typically assigned experimental or research projects, and when operating patrol-type aircraft conducted experiments with new or prototype ASW weapons and electronic equipment, etc., to determine their value and how they should be employed to the best advantage.)

Shown in December 1953 is BuNo 130286. This aircraft was built as a standard P5M-1 and was used to prototype the T-tail design of the P5M-2. (U.S. Navy photo)

Shown at NAS North Island in 1960, BuNo 130286 continued to operate as a P5M-1 with several fleet squadrons before being stricken from service in 1967. (Warren M. Bodie Collection)

1. Variable Incidence Homing Torpedo Launcher Evaluation. Both aircraft were equipped with special wing mounted torpedo launchers, which would permit launch of ASW homing torpedoes while on the water.

2. AN/AQS-6 (XN-2) Dipping Sonar Evaluation. Designated project OP/V245/J15-11. Raytheon designed and built seaplane sonar requiring extensive hull modification to accommodate the ten foot tall, 1,200 pound sonar transducer array. When stowed inside the aircraft, the transducer array collapsed to approximately half its operating diameter of fifty-two inches. Further modifications included the use of Curtis Electric propellers for precise thrust control during waterborne operation of the sonar. Fine tuning of the propellers was necessary to prevent tilting of the sonar transducer as the aircraft *hovered* over the sonar. This technique reduced forward motion of the aircraft to a minimum. To prevent overheating, sea water was pumped through special heat exchangers to cool engine oil during operation of the sonar equipment.

Following a series of product development tests, BuNo 140147 was re-configured for fleet assignment. This aircraft is known to have served with VP-40 and VP-50 during the Vietnam conflict, before being scrapped in Japan, in 1967. The disposition of BuNo 140146, following its product development testing, is not clear; however, it is thought to have returned to operational service with Atlantic based patrol squadrons as early as 1959.

BuNo 140146 is shown during testing of the Variable Incident Homing Torpedo in September 1957. Delivered to VX-1 in 1957, this aircraft was one of two production P5M-2s that was modified and used for special projects in the late 1950s. (U.S. Navy photo)

Shown undergoing modification by Martin, this SP-5B, BuNo 141258 was retrofitted with a J-60 jet-engine in 1964. Note the small rear hatch used to facilitate installation of the tail beaching gear. (U.S. Navy photo)

✪ P5M-2: BuNo 141258

In 1964, a single SP-5B *Marlin* was fitted with a 3,000-pound thrust jet-engine to evaluate performance enhancements, relative to a reduction in takeoff time and distance at increased gross weight, and for flight safety during single engine out conditions. Designed and proposed by Cdr. Richard A. Hoffman, USN, a test pilot with a Master's Degree in Seaplane Hydrodynamics stationed at NATC, Patuxent River, the design had the merits of being technically feasible, operationally desirable and economically practical.

The center-line-mounted installation of the jet - engine in the tail did not require extensive airframe modification since that area was originally designed for a twenty-millimeter tail turret which, although replaced by additional ASW equipment, weighed more than the proposed jet-engine. The center-line mounting minimized both longitudinal aerodynamic and asymmetric thrust control problems. Jet exhaust and wake disturbances were further avoided since they did not impinge on the aircraft's structure. The air intake for the jet was a reverse flow type mounted in the former tail gunner's window which was well outside the main spray envelope.

The small jet-engine was expected to produce 3,000 pounds of thrust, which at 161 miles per hour true airspeed, was approximately equal to 1,200 horsepower at sea level. It was estimated to reduce takeoff time from fifty-six seconds to thirty seconds and takeoff distance from 5,000 feet to 2,600 feet at an increased gross weight of 80,000 pounds. It was also expected to increase single engine rate of climb by up to 350 feet per minute.

Installed in place of the RT-533/APN-122 Doppler-Inertial Navigation antenna, the test aircraft was initially fitted with a Pratt & Whitney J-60 jet-engine and later upgraded to a General Electric J-85 jet in its final configuration.

Shown is BuNo 141258 at NATC, Patuxent River, during initial flight testing of jet modification by VX-1.
(U.S. Navy photo)

Modified by Martin in Baltimore, it was evaluated and tested at NATC, Patuxent River and VP-31, NAS North Island.

Although the engineering study and flight tests were completed, the project was not accepted by the Navy as a solution to prolong the aging design of the P5M. The project was terminated in 1966. Left in its modified state, the aircraft was removed from service and placed in storage at NAS North Island where it was eventually stricken from naval inventory.

✪ P5M-1: BuNo 124911 and 126490

In June 1953, the Flight Test Division of the Navy Department Board of Inspection and Survey (BIS) conducted the hydrodynamic characteristics phase of the service acceptance trials on the P5M-1 aircraft. Using two standard production Navy aircraft, BuNo 124911 and 126490, BIS pilots performed twenty-seven takeoff and landings in seas up to six feet while carrying gross weights from between 60,000 and 73,000 pounds.

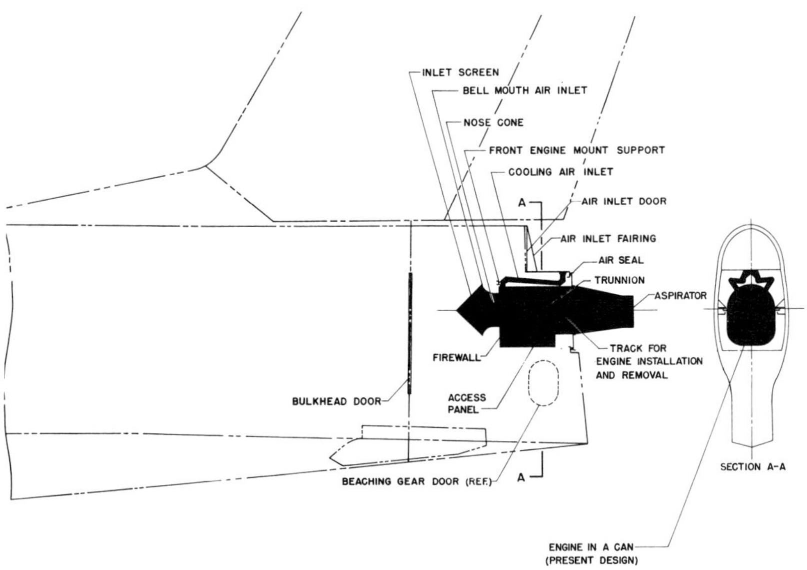

This drawing shows the initial configuration of the Pratt & Whitney, J-60-P-2 jet-engine that was installed in BuNo 141258 in 1964. (U.S. Navy)

BuNo 141258 was later upgraded with a General Electric J-85 jet-engine which included a revised exhaust cone, as shown in this 1966 photo taken at NAS North Island. (Clay Jansson photo)

Another feature of the J-85 modification to BuNo 141258 was the addition of a water separator, as shown here in 1966. The installation of the J-85 was the final configuration of the project.
(Clay Jansson photo)

Support Programs and Projects

Between 1955 and 1957, the Navy conducted a series of classified experiments to extend mobility concepts and boundaries of the seaplane beyond that of standard tender operations. The primary purpose of these experiments were in the Navy's anticipation of acceptance and delivery of the jet-powered Martin P6M *Seamaster*. Still in its developmental stages the P6M was not yet available and testing commenced using P5M-1 and P5M-2 *Marlins* from VP-44, VP-45, VP-46 and VP-56. The projects consisted of three main categories; shipboard dry-dock, floating docks and submarine refueling.

1. Shipboard dry-docking experiments were conducted using the highly modified LSD (Landing Ship Dock), USS *Ashland*, LSD 1. Having an overall length of 458 feet, the concept allowed for a 394 foot well deck that could be flooded to a depth of ten feet, accommodating three P5M series aircraft; however, for these experiments a temporary barrier was erected 130 feet from the stern tailgate.

   In performing these experiments, the stern, normally seventeen feet out of the water, was flooded to a depth of thirteen feet, giving the ship a three-to-five degree slant from bow to stern. With the ship at anchor and the tailgate lowered, the P5M, with the aid of an auxiliary craft, maneuvered head-long into position at the stern of the ship. With a bow line attached, the three sets of beaching gear were installed and the aircraft was drawn up the ramp and into the well. The stern tailgate was then closed and the ballast pumped out. The entire procedure took less than a half hour.

   Employed as a core development unit to improve water based aircraft servicing systems, the *Ashland* project was terminated with the demise of the P6M *Seamaster* program in 1959.

2. Floating Docking Systems. Used as a means of extending the serviceability of the seaplane on the water, these portable systems were used in conjunction with both the *Ashland* and *Guavina* projects. The simplest system was comprised of floating docks made from a series of inflatable rubber cubes, called "sealed bins," that could be deployed from the rear of a ship. When lashed together and partially filled with water, the bins submerged, with eighteen to twenty-four inches remaining above the water, serving as a pontoon bridge for personnel.

   The "U" dock system was designed primarily as a work station platform. It offered an alternative means of servicing the aircraft on the water to using conventional *sea-stands* or hoisting the aircraft aboard a tender. The docking system consisted of three, prefabricated sections joined together in a "U" shape. When fully assembled, the work station platform measured forty-three feet long and thirty-three feet wide and was supported by eighteen pneumatic rubber fenders.

3. Submarine Refueling. Commissioned in 1944, the Balao Class killer submarine, USS *Guavina*, AO(SS)-362, was the first of its kind modified for the refueling of seaplanes at sea. Converted and reclassified as an Oilier (AO) in the mid-1950s, the 311 foot long, diesel-powered submarine participated in numerous exercises with the P5M *Marlin*.

   Equipped with a sixteen foot plaster boat (powered by a twenty-horsepower motor), the submarine was capable of laying out a seadrome and standard mooring buoy for an aircraft. Once tended, refueling was accomplished using one or more of the following methods.

- Submarine surfaced, underway, aircraft alongside
- Submarine surfaced, at anchor, aircraft alongside
- Submarine surfaced, underway, aircraft astern
- Submarine submerged to sixty feet, under way, aircraft towed astern
- Submarine submerged to 200 feet, bottomed, aircraft moored on buoy

The *Guavina* had the capability of supplying the aircraft with spare parts, JATO bottles, hydraulic and lubrication oils, flight rations, potable water and personal items for the crew. Not a substitute for the seaplane tender, the primary mission of the submarine was to provide an undectable seadrome, effectively extending the operating range of the seaplane.

In 1957, a series of docking experiments were performed with the USS *Ashland*, LSD-1, using both the P5M-1 & P5M-2. Shown here is a P5M-1 from VP-44 being drawn into the flooded stern of the ship. (U.S. Navy photo)

This P5M-2 from VP-56, BuNo 135529, is shown on its beaching gear, completely inside the stern well of the USS *Ashland*. (U.S. Navy photo)

Once the P5M was completely inside the well of the ship, the stern tailgate was closed and the ballast pumped out, raising the stern to normal draft level as shown. (U.S. Navy photo)

Shown is VP-44's P5M-1 demonstrating the use of a floating dock towed from the stern of the USS *Ashland.* (U.S. Navy photo)

The inflatable rubber cubes shown here were called *sealed bins.* These could be lashed together to serve as a pontoon bridge from the ship to the aircraft. (U.S. Navy photo)

This prefabricated, floating U-dock provided a sturdy work platform around the aircraft that could be used for various projects, including loading heavy aerial mines as shown. (U.S. Navy photo)

The USS *Guavina*, AO(SS)-362, was the only submarine converted for tending seaplanes at sea. (S. Goodman Collection)

4. The Curtis class seaplane tender, USS *Albemarle,* was projected to become Naval Air's answer to logistic support for the future jet-powered P6M *Seamaster.* Commissioned in 1940, the 537 foot long, eighty-seven foot wide tender remained in active service until deactivation in August 1950. Having undergone a complete refurbishment and conversion to handle the P5M and P6M aircraft, the tender was re-commissioned on October 21, 1957.

Following shakedown and Operational Readiness Inspection (ORI) in Guantanamo Bay, Cuba in January 1958, the *Albemarle* teamed up in March with four Atlantic based *Marlin* squadrons participating in competitive fleet exercises entitled *Operation Springboard.* With the demise of the P6M project in 1959, the tender was reclassified and renamed, T-ARVH-1, and assigned to the Military Sealift Command for use as a floating repair facility by the U.S. Army. It was stricken from naval service on December 31, 1974.

Shown during flight testing in March 1959 is BuNo 145876, the first production P6M-2. (Martin Aircraft Company)

The Martin P5M
Marlin

Chapter Five

The Interim Years

Just as they were in 1940 when they took delivery of the first PBM *Mariner*, the Navy was ready for the *Marlin* when it was introduced to the fleet in 1952. Met with overwhelming approval, squadrons set out to prove the *Marlin* a success in the fleet. Working with Martin's technical representatives in the field, squadrons played an active role in improving maintenance techniques and documentation for the new aircraft.

As early as 1949, the Coast Guard had initiated a series of studies with the PBM-5 to determine the best way to land a seaplane or to ditch a fixed wing aircraft in the open ocean. Under the direction of Cdr. Donald B. MacDiarmid, USCG, the joint project with the Navy and the Martin Aircraft Company resulted in establishing base-line procedures for open-ocean operations that would be carried over to the P5M.

Most, if not all the squadrons flying the P5M in the early 1950s, were accustomed to either the PBY *Catalina* or PBM *Mariner* flying boats prior to obtaining the *Marlin*. Now in operation with several squadrons throughout the country, the *Marlin* was put through a myriad of tests at the squadron level. A completely new aircraft, the advanced design of the P5M lacked the benefit of proven performance and handling characteristics on the water. Unsure of the *Marlin's* limitations, squadrons performed routine hydrodynamic tests with the new aircraft in various sea-states, honing their skills in open water with the new long afterbody hull and experimenting with the use of JATO. Many of these tests were performed in 1953 and 1954 with VP-40, 44, 46, 48 and the U.S. Coast Guard.

In the tranquil period that followed the end of the Korean War in July 1953, the Navy's growing number of seaplane squadrons flying the P5M found themselves dispersed throughout the world in such places as Iwakuni, Japan; Sangley Point, the Philippines; and the island of Bermuda in the Caribbean. Whether rotating six-month deployments or operating in a permanently deployed status, the squadrons performed ASW exercises, ocean surveillance and air-sea rescue operations.

Incidentally, the seaplane facilities in Iwakuni were originally built by the Japanese as their main naval air training base. During World War Two, the Japanese used the air-station as a Kamikaze base. Still visible during the Korean War was the crane used to lift the suicide bombers from their dollies and the mechanism used to launch the aircraft on their one-way flight. It's reported that at the end of the war twenty-eight Japanese Kamikaze pilots marched to the crane where they committed hara-kiri. Following the war, the base was used by the British before being taken over by the U.S. Air Force. Eventually, a Marine Corps Air Station (MCAS) was established at the facility.

A formation of early P5M-1s from VP-56 is shown over Norfolk circa 1954. The aircraft in the foreground, EH-9, is BuNo 130273. (U.S. Navy photo)

The new seaplane was yet to be proven with the seaplane tenders, whose crews were unfamiliar with the handling characteristics of the larger flying boat. Commencing on July 8, 1953, VP-40 teamed-up with the seaplane tender, USS *Salisbury Sound* (AV-13), off the coast of California for a week long exercise to evaluate the aircraft under operational conditions. Accustomed to the earlier PBM, the seaplane and tender crews performed an exhausting series of tests with the aircraft including towing, hoisting and fueling of the big *Marlins*. Overcoming their unfamiliarity with the new aircraft, the exercise was completed and documented for future use in training shipboard and squadron personnel in the handling of the *Marlin*.

In 1953, the Coast Guard placed an initial order through the Navy's contract with Martin for seven P5M-1s to supplement their complement of UF-1 *Albatross* seaplanes. Assigned a "G" suffix on their primary designator (P5M-1G), they were configured without the Navy's advanced ASW electronics and defensive armament equipment and used primarily for long-range air-sea rescue. Prior to receiving their first *Marlin* in early 1954, Coast Guard crews attended a two week ground school on the new aircraft at Martin in Baltimore.

The Coast Guard used the same twenty-four month overhaul cycle for their aircraft as did the Navy; however, unlike their Grumman *Albatross* seaplanes which could be overhauled in a period of three months by Coast Guard personnel, the new P5Ms were overhauled by the Navy and often took up to six-months to complete. Operating costs per hour to maintain the complex P5M were high, given the lack of actual use by the Coast Guard. Eventually the high cost became the determining factor in the decision to return them to the Navy.

The Navy took the *Marlin* on its first operational commitment outside the boundaries of the continental United States in early 1954. In January, just ten months after receiving the new aircraft, VP-40 departed NAS North Island with their complement of twelve *Marlins* for a six-month operational deployment in the Western Pacific (WestPac).

While operating as part of the Formosa Patrol Force, VP-40 was based at Naval Station Sangley Point, Republic of the Philippines. From there, operations were extended to Buckner Bay, Okinawa, Japan, and the Pescadore Islands, where they patrolled the Formosa Strait and South China Sea.

Although faced with a heavy maintenance load, due in part to their unfamiliarity with the new aircraft, acute shortage of spare parts, and poor maintenance facilities, VP-40 flew over 6,000 flight hours during this first deployment with the *Marlin*.

Getting hosed down on the ramp at NS Coco Solo, Panama is VP-45's BuNo 135461 in 1955. A predominantly saltwater aircraft, the *Marlin* required a fresh water washdown after each flight. (Al Raithel photo)

Shown in the foreground of this 1951 photo is the original ramp and crane used for launching suicide planes from the post-war allied air station at Iwakuni. The station served as a *Kamikaze* base during WW II.
(Charles Highfill photo)

Pictured in storage, at NAS North Island in 1965, is a P5M-1 from VP-30, BuNo 130301.
(Clay Jansson photo)

This November 1953 photo shows a class of USCG pilots attending ground school training on the P5M-1 at Martin.
(Courtesy of RAdm. J. Steele, USCG Ret.)

Shown is the first USCG P5M-1G, s/n 1284, leaving the Martin paint hanger at Strawberry Point on October 22, 1953.
(Martin Aircraft Company)

This photo of VP-42's BuNo 127699 was taken in 1959. This aircraft, which completed the Modification Program, displays the later tail code, RB, introduced in September 1958. (U.S. Navy photo)

In this 1956 photo, a P5M-2, BuNo 135514, from VP-42, is about to be brought up the ramp at Sangley Point. Note the tail turret and the swimmer in the water to assist with the tail beaching gear. (U.S. Navy photo)

While deployed to the Mediterranean in 1954, VP-44 performed operational exercises with the P5M-1 and the USS *Currituck*, including hoisting as shown here with CC-11. (U.S. Navy photo)

A PBM-5S2 from VP-50, BuNo 98609, is shown at NAS Alameda in 1955. In July of the following year, VP-50 officially retired the *Mariner* from fleet service. (Warren M. Bodie Collection)

Averaging 942 hours per month, with an average of eight planes in commission, the squadron was able to meet all operational commitments, achieving an overall availability of seventy-two percent.

As with the early versions of the PBM, the P5M also suffered from an unusual number of low-time engine failures on their initial production engines. The problem was caused in part by metal contamination in the oil of the early R-3350-30WA engines. This resulted in frequent replacement of the oil cooler, or in the worst case, necessitated changing the entire engine. This problem was later remedied by the manufacturer (Curtiss-Wright) through a series of engineering upgrades and revised maintenance procedures. Eventually, the -30WA was removed altogether from the P5M-1 and replaced by one of several upgraded versions including the -32WA, -36W and -36WA.

Later that year, VP-44 deployed their *Marlins* to the Mediterranean for a fifty-five day period. Depending entirely on support from the seaplane tender USS *Currituck* (AV-7), squadron aircraft logged in excess of over 400 flight hours performing routine ASW and training missions without incident.

Not all seaplane squadrons in the early 1950s transitioned to the P5M-1 initially, as the Navy was also still operating the PBM *Mariner*. Some squadrons retained their PBMs and transitioned directly to the later model, the P5M-2. An example of this was VP-47, who had flown the PBM since 1944, before making the transition to the P5M-2, in 1954. Patrol Squadron Fifty started out flying the PBM and transitioned directly to the P5M-2 in 1956. Training squadrons continued to use the PBM-5 *Mariner* for navigational training until 1958, retiring them in favor of the newer *Marlin*. In late 1958, ATU-501, NAS Corpus Christi, Texas officially retired their last PBM-5 (BuNo 84685).

While based in Bermuda, VP-49's Crew Three and P5M-1, BuNo 127706, participated in the making of the 1955 film, *Look to the Sea*. (U.S. Navy photo)

This P5M-2G, s/n 1318, shown in San Diego, was the second of four delivered to the USCG. (Clay Jansson photo)

Still in Coast Guard markings, at NAS North Island in 1965, this VT-31 P5M-1T, BuNo 149825, was originally USCG P5M-1G, s/n 1284.
(Clay Jansson photo)

Some *Marlins,* used for training, were painted with bright orange on the nose and vertical stabilizer, as shown on this P5M-2, BuNo 147931, at NAS North Island in 1962. Commissioned in 1960, VP-31 was the Combat Replacement Patrol Squadron for the Pacific.
(Clay Jansson photo)

By late 1954, the first of the newer P5M-2 models started to reach the fleet. As the number of seaplane squadrons began to increase, it was a common occurrence for squadrons to rotate use of both the dash-one and dash-two models, often flying both. Similar in design and performance, pilots and crews soon developed a proficiency in each. It was not until the early 1960s that the dash-one model was phased out of the naval inventory.

In 1955, VP-49 helped produce a promotional film on the P5M-1 *Marlin* in conjunction with the Martin Aircraft Company. Filmed in Bermuda, this fourteen minute color documentary, titled *Look to the Sea*, promoted the role of the naval seaplane as an advanced ASW patrol aircraft.

Aerojet General Corporation produced a series of twenty-six half-hour films entitled *On Guard* in 1956. Their purpose was to demonstrate the Armed Forces capabilities and state of combat readiness. For a segment in the series entitled *Hunter Killer*, several P5M squadrons stationed at NAS North Island participated in the making of the film. This segment showed how naval surface and air units would team up to seek out and destroy enemy submarines in the event of hostilities.

The Coast Guard increased the number of *Marlins* in their inventory with an additional order for four of the later dash-two models in 1956. Operating from stations in Elizabeth City, North Carolina; St. Petersburg, Florida; San Francisco and San Diego, California, the Coast Guard continued to operate both the P5M-1G and the P5M-2G simultaneously. With a total of only eleven *Marlins* available for long-range air-sea rescue operations throughout the country, the six-month downtime required for overhaul and the rising maintenance costs per flight hour were quickly rendering them cost prohibitive to operate. Seizing

A P5M-1 from VP-42, BuNo 130295, taxies towards the seaplane ramp at NAS North Island in late 1958. (Clay Jansson photo)

Shown being hoisted aboard the tender, USS *Salisbury Sound*, is this P5M-2 from VP-48, BuNo 135504. (U.S. Navy photo)

A P5M-2, from VP-49, makes the buoy while operating in Bermuda in 1962. (U.S. Navy photo)

Shown is VP-50's SG-7, BuNo 135496, being brought up the ramp at NAS North Island after a local training flight in 1965. (U.S. Navy photo)

VP-40's QE-3, BuNo 135475, taxies past the USS *Currituck,* AV-7, in 1962. This P5M-2 was the first accepted by the Navy in 1954. (U.S. Navy photo via Frank Schneider)

While the crew *slept* on buoy watch, this P5M-2 from VP-40, ruptured a wing-tip float and tipped over in Buckner Bay, Okinawa in 1963. This aircraft was salvaged, repaired and returned to service. (U.S. Navy photo)

an opportunity to acquire several low-time HU-16 *Albatross* amphibians from the U.S. Air Force, the Coast Guard returned all eleven *Marlins* to the Navy in 1960.

Redesignated P5M-1T, these brightly colored Coast Guard aircraft were assigned current Navy Bureau Numbers (BuNo) and used for fleet training of pilots and crew at VT-31, NAS Corpus Christi, and VP-31, NAS North Island until being phased out of the Navy's inventory in the early 1960s in favor of the later version, the P5M-2.

As a useful aid to flying the P5M series aircraft, personnel from VP-49 conceived and prepared a series of charts called the *Howgozit* in 1957. Used for both normal and single engine flight, the *Howgozit* was developed to provide an accurate means of cruise control during standard operational and extended flights. The charts were a method of checking fuel consumed, verses distance traveled and time elapsed, and were of particular importance during single-engine operation. (This was just another example of the degree of squadron involvement in supporting and promoting the success of the P5M.)

A step-by-step instruction booklet (Martin CSR-401) was published, providing information on how to prepare and execute the *Howgozit* for each model of the *Marlin*. When properly preflighted, a flight could be efficiently conducted with accurate in-flight checks, checking the *actual* results against *predicted* results, when using the *Howgozit* chart. At a moment's glance, one could easily see how the flight was progressing with respect to (1) fuel consumed, (2) distance traveled, (3) elapsed time, (4) existing weight, (5) amount of time and fuel required to return—single engine, if necessary—and (6) quantity of weight needed to be jettisoned.

In 1957, VP-50 set a new speed record for trans-Pacific flight in a P5M-2. Departing the island of Oahu, Hawaii, on February 24 for a flight to NAS Alameda, Crew Eleven in BuNo 135477 completed the 2,140 mile flight in 9 hours and 20 minutes. Apparently this is believed to be a record for a twin-engine seaplane in the *Marlin* class.

In early 1958, several squadrons participated in *Operation Springboard* with the seaplane tender USS *Albemarle*, in San Juan, Puerto Rico. During this two week exercise in March, VP-45 flew its twelve *Marlins* to a *Springboard* record of 852 hours. Their success in establishing this record was directly attributed to the combined team work of squadron and shipboard personnel and that of the Martin technical representatives assigned to the project. The total time flown by the four seaplane squadrons, competing in this two week exercise, amounted to over 2,800 hours.

- -

A French P5M-2, BuNo 146440, shown in 1958. The French painted their *Marlins* a gray-green and white with colorful tail markings.
(Martin Aircraft Company)

Shown at Norfolk in 1959 is BuNo 146442. The *Marlin* was used extensively by the French Naval Air Force in West Africa through 1964. (Martin Aircraft Company)

Patrol Squadron Forty-Five was not the only squadron to set a record during this deployment exercise. Four-hundred miles out over the Atlantic on their return trip to NAS Norfolk, VP-44's number eight P5M-2 (LM-8, BuNo 135481) developed a serious fire in the starboard engine. Flying at an altitude of 8,000 feet, the aircraft began to drop quickly. The aircraft's commander, Lt. Raymond W. Myers, made a decision to ditch at sea, rather than order his crew to bail out of the crippled aircraft. Within minutes the aircraft was on the water. Immediately jettisoning the auxiliary bomb bay fuel tank in the starboard nacelle and feathering the damaged engine, the flaming aircraft made an uneventful open sea landing. Safely on the water, the crew formed a human chain to pass sea water up to the burning wing, successfully extinguishing the blaze.

Using only the port engine, the aircraft, with its crew of thirteen, taxied nearly 200 miles to Grand Turk Island, B.W.I., to rendezvous with the tender, *Albemarle*, in order to assess the damage. Inspection of the aircraft revealed serious metal damage to the wing and probable damage to the main spar, apparent by a drop of the starboard engine of nearly one foot. The damage was much too serious to be repaired by tender personnel.

Accompanied by the *Albemarle*, the aircraft proceeded to taxi under its own power to the nearest port with facilities to repair the damage; however, the nearest port was Guantanamo Bay, Cuba, some 550 miles away.

Laughingly dubbed the *Caribbean Cruiser* by its crew, LM-8 set out on what was to become the longest known water taxi by a naval seaplane. Averaging ten knots on the water, both aircraft and crew arrived safely at their destination at the Naval Base, Guantanamo Bay, in less than three days after their emergency landing. To be exact, it took two days, 6 hours and 52 minutes to cover the 550 miles—an undisputed record.

- -

The Martin P5M was without a doubt the most modern production medium-range ASW seaplane in its day and had no competition at the time. Up until the late 1950s, the French were still operating World War Two vintage Short *Sunderland* seaplanes from their base in Dakar, West Africa. Seeking to replace these aging flying boats with a more modern seaplane, the French were unable to find a suitable replacement on the European market and turned their attention to the United States and the P5M. In early 1959, the U.S. supplied ten P5M-2 *Marlins* to France on a lend-lease agreement under the Military Assistance Program.

While in flight in 1962, a single JATO bottle can be seen attached to the waist hatch on VP-49's LP-5, BuNo 147932.
(U.S. Navy photo)

During tender exercises in 1962, VP-40's Crew Three, aboard BuNo 135475, prepares the aircraft for hoisting aboard the USS *Currituck*. Notice the launch, with a cable attached to the rear of the *Marlin,* controlling directional stability, while the hoist is being affixed to the aircraft.
(U.S. Navy photo)

While deployed to NS Sangley Point in 1965, VP-40's BuNo 135476 is shown stored at NAS North Island.
(U.S. Navy photo)

This P5M-1 from VP-40, BuNo 126501, was the first *Marlin* to undergo refurbishment by Shin Meiwa in July 1960.
(Shin Meiwa Industries)

Aircraft were delivered to the Japanese facility in various states of repair, such as this SP-5B minus a tail in February 1966. (Shin Meiwa Industries)

During formal ceremonies in Baltimore on January 9, 1959, the French accepted delivery of their first *Marlin*. Destined to replace the *Sunderland* MR5s which had been in service at Dakar, Bel-Air since 1943, the French looked forward to the newer, more modern flying boat. Following receipt of their ten new *Marlins*, the French crews and support personnel proceeded to spend the next four months training with their aircraft at NAS Norfolk and NAS Corpus Christi. By the beginning of May, the French had completed their training and made preparations to return to their home-port at Dakar.

In early May, the first of the French *Marlins* ventured out over the South Atlantic, flying the same route as that of famous French aviator, Jean Mermoz, who pioneered these vast stretches of open sea in the early hydroplanes during the 1930s. Leaving the coastal edge of the U.S. behind, the attractively painted flight of *Marlins* flew onto Guantanamo, Cuba on the first leg of their journey.

Unlike their American counterparts, the French navy painted their *Marlins* a pleasing gray-green color with white underpinnings and a brightly painted rudder, displaying the French tri-colors and the unit insignia, the *Southern Cross*. Nicknamed *penguins* by the French, they were the most attractively finished of all *Marlins* flown.

Leaving Cuba, their journey continued on to Port of Spain, Trinidad, then down the coast of Venezuela and into the mouth of the Amazon before making their third stop at Belem, Brazil. Leaving the Brazilian coast behind, they headed on to Natal, South Africa for the fourth and final stop before reaching their destination. The last leg of the journey was nearly 2,000 miles long. They completed this segment in ten hours without incident, arriving in Dakar on May 8, 1959. Three days later on May 11, an open house was held by the 27F Flotilla to introduce the new Martin flying boat to squadron personnel based at the Bel-Air facility.

Although old and tired, the *Sunderland* enjoyed a reputation for durability and ease of maintenance, and initially, the new *Marlins* suffered from a low serviceability rate. This was due in part to heavy flight commitments, high humidity operation and the complexity of new equipment. Adjusting to the new aircraft, this problem was soon overcome and the work became routine.

Assigned to the French Naval Air Force, Flotilla 27F based at Dakar, the *Marlins* flew exhausting patrols of up to fourteen hours endurance in the hot, humid African climate. The high humidity was torturous on the aircraft and often required a complete fresh water wash-down after each flight. Routine maintenance was carried out in Dakar, while heavier maintenance was performed at St. Mandrier near Toulon, France, some

This intact SP-5B from VP-47, BuNo 135517, is shown being unloaded at the Konan facility in December 1964.
(Shin Meiwa Industries)

eleven hours away. Like their American counterparts, the French eventually changed the colorful paint scheme on their *Marlins* to include a white reflective surface on the top half of the fuselage. This helped to reduce the solar effect of heat buildup inside the aircraft.

While in the possession of the French, the *Marlins* enjoyed close to an accident free record. The only accident recorded was a ground incident involving BuNo 146445 and 147539 in 1961; however, both aircraft were quickly repaired and returned to service.

The French retained the aircraft in operational use until August 1964. Based mainly on a decision by the French government to withdraw its Atlantic naval forces from NATO in 1963, the 27F Flotilla was dissolved, and all ten aircraft were returned to the U.S. Navy in Baltimore by October 1, 1964.

As the number of squadrons flying the *Marlin* in the Pacific increased, it became necessary for the Navy to establish a major repair facility in the area. In early 1960, Shin Meiwa Industries located in Konan, Japan was awarded a contract by the U.S. Navy to serve as a repair facility for all P5M aircraft deployed in the Pacific. They also provided storage for replenishment aircraft and acted as a disposal facility for those aircraft stricken from service as scrap.

The first aircraft to undergo rework by Shin Meiwa was a P5M-1 from VP-40 based in the Philippines. Flown to the Konan facility from NS Sangley Point, BuNo 126501 arrived on July 5, 1960. On September 2, 1960, the first scheduled refurbishment was completed. Recognized as a significant achievement by the Japanese facility, it was acknowledged by a formal ceremony attended by company and military personnel alike.

In all, 128 P5M *Marlins* were sent to the facility at Konan for rework between 1960 and 1967. They arrived either by air or aboard ship, depending on condition and degree of repairs required. Typical repairs performed at the facility consisted of major corrosion control, structural repairs and general refurbishment. The facility also had the ability to completely overhaul and refurbish the P5M and this was often the case when they returned from Konan.

While permanently based in the Philippines, VP-40 took part in a significant air-sea rescue operation with the P5M-1. On July 14, 1960, a Northwest Airlines DC-7C was forced to ditch at sea off the Luzon coast of the Philippines, following two serious engine failures. Carrying a total of fifty-eight passengers and crew, the aircraft made a successful ditching with the loss of only one passenger. Under the direction of Cdr. Irving G. "Moe" Cockroft, commanding officer, flight crews from VP-40 worked in concert with those from the Air Force, flying SA-16 *Albatross* seaplanes, performing a successful rescue of all passengers. For their accomplishments during this rescue, those Navy pilots involved were awarded the Air Medal, with their crews receiving the Navy's Commendation Medal.

Over the years, many *Marlins* were stricken from use due to irreparable structural damage and because of accidents, but the majority of those stricken from active service prematurely fell victim to the elements of the weather. As a flying boat, the 30 foot high, 40 ton *Marlin* had difficulty fitting in most hangers. As a result, they were often stored outside or moored on a buoy, unprotected from the elements of their surroundings. While operating from remote areas in the Western Pacific, squadrons routinely encountered typhoons; violent tropical storms that could shred and destroy the unprotected aircraft. Each situation was different, requiring a decision whether or not to evacuate the aircraft to a safe place or tie them down and ride out the storm.

It was during a deployment to the Philippines that one squadron faced three such decisions in a six-month period. Electing to avoid the first two typhoons, the squadron evacuated their P5M-1s to the tiny town of Puerto Princessa, located on the island of Palawan, some 500 miles south of NS Sangley Point. Later that year, a detachment operating out of Buckner Bay, Okinawa was forced to ride out a third typhoon while moored in the bay. Of the many incidents of this type, this particular story notes with grim reality the personal sacrifice experienced by a *Marlin* crew to save their aircraft from destruction.

- -

According to the *Canacao Clipper* (Sangley Point Newspaper) dated August 24, 1956; "Ltjg. Marvin Register of VP-40 was rescued by an Okinawa fisherman after hours of clinging to a tiny coral reef as winds up to 100 knots howled around him. This completed the successful rescue of all five VP-40 airmen who were dumped into the typhoon-whipped Pacific after their P5M-1 was torn loose from its mooring and blown onto a reef."

Part of a three plane detachment operating in the Okinawa area, the crew of five boarded their aircraft (CA-1, BuNo 127700) on August 14 to stand watch during the eve of typhoon "Babs." By the next morning the sea-state was too rough to takeoff or for anyone else to come aboard. A decision was made to try to ride out the impending storm. When the aircraft broke loose from its mooring, the men stayed aboard in an attempt to save it; however, after taxiing around for almost ten hours, the aircraft was blown into the rocks. Separated in their attempt to abandon the damaged aircraft, three of the four enlisted men aboard paddled to safety in a rubber raft just as the aircraft started to break apart. Blown ashore by the severe winds, they were later rescued by a team of Marines from the base. The fourth enlisted crew

Shown at NAS North Island in 1960 is VP-47's RD-4, BuNo 140141. Notice the complete BuNo on the tail and the Efficiency "E" on the bow. (Clay Jansson photo)

Shown at NAS North Island, this P5M-1 from VP-46, BuNo 135473, was the last P5M-1 manufactured in the series. (Clay Jansson photo)

This excellent view of VP-45's LN-8, BuNo 135505, was taken in Bermuda in 1962. Note the *chevron* painted below the cockpit and the markings on the aft fuselage that say *Task Group Delta*. VP-45 was the last Atlantic squadron to fly the *Marlin* in 1963. (U.S. Navy photo)

member, initially listed as missing, was later rescued from a nearby island by a search party of Marines from the 1st Marine AmTrac Battalion.

The first attempt to rescue Register failed as three-foot high waves forced the rescuers back. Joined by seven Okinawins in a fishing boat, two Marines headed back out to the reef in a second attempt to rescue the pilot. Working as close to shore as possible, one of the Okinawins swam the remaining thirty feet to reach the stranded airman. Safely on the boat, he was returned to shore. In all, none of the men were seriously injured, though several suffered numerous cuts and bruises from their ordeal on the reef.

- -

In June 1957, *Naval Aviation News* reported another weather related story about a *Marlin* that was forced to land in a reservoir due to weather and was eventually beached on a mud flat to prevent sinking from damage incurred during an aborted takeoff attempt. While based at NAS Whidbey Island, a P5M-2 belonging to VP-50 (SE-9, BuNo 135534) departed the local seadrome on a routine instrument flight in early January. Not long into the flight, the weather worsened in the Whidbey area with heavy snow and diminishing conditions. With little more than five hours of fuel remaining, the crew selected an alternative landing site and changed their flight plan. Heading south towards Alameda, near San Francisco, the *Marlin* encountered severe icing conditions and the crew was forced to alter their flight plan once again.

Now flying at a lower altitude to reduce the impending ice formations, the *Marlin* was running low on fuel. Noting a fairly large reservoir just ten miles from the Eugene Municipal Airport, the crew opted to put the big seaplane down. Landing without incident, the Navy worked with U.S.

The SP-5B in this 1965 photo, BuNo 147932, was among the last P5M-2s delivered to the Navy, excluding those sent to France.
(Clay Jansson photo)

Army Corps of Engineers to refuel and prepare the aircraft for a return flight to Whidbey when the weather improved; however, during pre-flight checks the aircraft hit a submerged obstacle and ruptured the hull. In order to prevent sinking or any additional damage from occurring, the aircraft was intentionally beached on a mud flat.

It took more than a week to effect repairs to the damaged hull. Working in the cold, wet mud, maintenance crews repaired the underside of the hull and water damaged electrical components. Once the repairs were completed, it took the combined effort of both the Portland Water District and the Army Corps of Engineers to adjust the water level in the reservoir to a depth suitable for the 78,000 pound aircraft to takeoff. Using a sealane of 4,500 feet, with a minimum depth of nine feet, VP-50's *Marlin* successfully departed the lake with the help of JATO, nearly three weeks after its unfortunate landing.

- -

The military's aircraft designator system changed in 1962 as part of a joint standardization program between the services. Adopting a system similar to one in use by the U.S. Air Force, the Navy's designation for the P5M was shortened to the P-5. Individual models were re-classified as P-5A and P-5B, and those having received advanced ASW upgrades were further identified by an "S" prefix. Using this new system, the P5M-1S was changed to SP-5A and the P5M-2S to the SP-5B.

By the time the new designator system went into effect, the early SP-5A models had already been retired from fleet service in favor of the newer SP-5B. However, training squadrons continued to fly the SP-5A until late 1963, at which time the model was retired from naval service completely.

When production of the P5M series concluded abruptly in 1960, a political decision was made to place the future of the seaplane in its military role on hold indefinitely in favor of more advanced land-based aircraft. Unsure of the future of the seaplane, the Navy initiated a program to reduce its number of fleet squadrons flying the *Marlin*. Making the transition to alternative patrol aircraft, the number of seaplane squadrons began to dwindle. By 1963, the Navy's complement of seaplane squadrons operating the SP-5B had been reduced to one Atlantic squadron and several Pacific squadrons. In December 1963, VP-45 made the transition to the P-3 *Orion*, leaving only four West Coast squadrons still flying the *Marlin*.

The Martin P5M
Marlin

Chapter Six

SEATO Operations

The South East Asia Treaty was signed in Manila on September 8, 1954, giving birth to an organization known as SEATO. It was created by its eight member nations to counter the threat of communist aggression in the area and to ensure a condition of security necessary for peaceful progress and development. Member nations included Australia, France, New Zealand, Pakistan, Philippines, Thailand, United Kingdom and the United States.

Basically, SEATO was a state department organization governed by the ministers of member nations with no actual military forces assigned; however, many maritime, air, ground and sea-land exercises were carried out to develop a military organization. Typically held in the Philippines, SEATO was among the largest of joint fleet exercises in the Western Pacific. Over twenty of these exercises were carried out by member nations between 1956 and 1967.

Participating U.S. Navy squadrons, flying the *Marlin,* were often called upon to serve as host squadron for those from other member nations. At each event, the Philippine skies were full of aircraft complementary of the vintage P5M, including the Royal New Zealand Air Force (RNZAF), Short *Sunderland;* Royal Australian Air Force (RAAF), Lockheed P-2H *Neptune;* and British Royal Air Force (RAF), Avro *Shackleton.* (As a point of interest, time expired for these accomplished veterans of the skies, for they too have been retired by their respective nations in favor of newer generation aircraft.)

During SEATO exercise *Sea Serpent,* VP-40's Cdr. John R. Johnson (right) was host to RNZAF Wing Cdr. John L. Scott (center). (John R. Johnson Collection)

In this 1963 photo, VP-40's QE-1, BuNo 135530, leads a flyover of SEATO participants at Sangley Point, during exercise *Sea Serpent.* (U.S. Navy photo via John R. Johnson)

The last SEATO exercise, employing the SP-5B, was conducted in the Philippines on March 13-16, 1967. Participating in this last exercise termed "SIYASAT," *Marlins* from VP-40 performed with aircraft from Australia, Thailand, United Kingdom and the Philippines.

Present at many of these exercises were the *Sunderland* GR5 seaplanes of the RNZAF. Unlike their American counterparts, these majestic four-engine flying boats were painted completely white and practically lived in the water, removed only for major rework. Not based locally in the Pacific, the New Zealand seaplane crews ferried their aircraft over 4,600 miles to participate in the exercises. Enjoying a reputation for reliability, the *Sunderland* first entered service around 1937 and quickly became the backbone of the British flying boat fleet during World War Two.

Another familiar participant in SEATO operations was the British MR-2 *Shackleton*. This post-war, medium-range bomber was a direct descendent of the legendary war-time Avro *Lancaster*. Powered by four Rolls-Royce Griffon, 2,500 horsepower, in-line, twelve cylinder, water-cooled engines, the vintage bomber was an impressive sight and easily distinguishable by its unique six-bladed counter-rotating propellers.

The Short *Sunderland* was a familiar sight during SEATO operations. Squadrons often ferried these aircraft over 4,600 miles to participate in the Philippines' exercises, such as NZ4115 shown in 1966. (B. Deatrick photo)

Also a familiar participant in SEATO was the British Avro *Shackleton* with its unique counter-rotating propellers. Notice the cockpit and wheel covers used to protect the vintage 1950s components in this 1966 photo taken at Sangley Point. (Author's Collection)

The Martin P5M
Marlin

Chapter Seven

The Final Years

By the time the United States entered the Vietnam conflict, plans were already in process to retire the naval seaplane. As hostilities increased in South East Asia, the Navy saw a need to put the big Martin SP-5B *Marlin* to the test. This was the first time, since the Korean War, that advance-based seaplane-tender operations were used under combat conditions and a first for the *Marlin*.

It was mid-1964 when communist activities started to escalate in the Tonkin Gulf, and an operation, known as *Market Time*, was formed to blockade the infiltration of arms and munitions into South Vietnam by sea. Scheduled for retirement in favor of the land-based P-3 *Orion*, the life of the naval seaplane was extended, and the three remaining *Marlin* squadrons, VP-40, VP-48 and VP-50, prepared for the most rigorous commitment of their career.

In July 1964, VP-50 was in the process of changing home-port to NAS North Island from MCAS Iwakuni, Japan, where they had been based since 1960. Leaving their *Marlins* behind in Iwakuni for disposal, the squadron was enroute to San Diego, via Navy transports, where they were scheduled to transition to the P-3. Patrol Squadron Forty-Eight was deployed to NS Sangley Point, where they had relieved VP-40 in March 1964. With operation *Market Time* in full swing, VP-48 sent crews to Japan to retrieve the additional aircraft left behind by VP-50. In turn, Patrol Squadron Fifty delayed their transition to the P-3 and returned to a rotational six-month deployment cycle, flying the *Marlins*.

Patrol Squadron Fifty was the first all reserve squadron to be activated in support of the Korean War. Based in Seattle, Washington, flying the PBY-5A *Catalina*, the squadron was operating under the designation of VP-892. Called into active service on August 4, 1950, the squadron made the transition to the PBM-5 *Mariner* and commenced deployment to Japan on December 18, 1950. On February 4, 1953, the squadron was designated VP-50. Continuing to fly the PBM-5S2, they transitioned to the P5M-2 in 1956. In 1967, the squadron made the transition to the P-3 *Orion* and changed their home-port to NAS Moffett Field. Between entering active service in 1950, and their decommissioning in June 1992, VP-50 experienced many changes in home-port, including Iwakuni, Japan; North Island, Alameda and Moffett Field, California; and Whidbey Island, Washington state.

Like their sister squadron VP-50, VP-48 was also a reserve squadron that was activated in support of the Korean War. Based at Grosse Isle, Michigan, under the designation of VP-731, the squadron was called to active duty on October 3, 1950. Assigned to NAS North Island, flying the PBM-5S *Mariner*, the squadron completed two tours in support of the Korean War before being

Shown during a training exercise at NAS North Island in 1965 is VP-50's SG-9, BuNo 147932, launching an aerial torpedo. (U.S. Navy photo)

On station in Vietnam in 1966, VP-50's BuNo 140148, is shown aboard the USS *Salisbury Sound* at Cam Ranh Bay.
(U.S. Navy photo)

VP-48's Crew Seven, with BuNo 141256, is shown on the ramp at Iwakuni in 1964 as their SP-5B is fueled for a patrol in the Tonkin Gulf.
(Ed Calloway photo)

designated VP-48 on February 4, 1953. The squadron made the transition to the P5M-1 in June 1954 and the P5M-2 in mid-1960.

Upon entering the Vietnam conflict in 1964, VP-48 established a record for a seaplane squadron by accumulating 1,585 flight hours in a single month, defending operation *Market Time*. Flying around the clock, the squadron averaged two planes airborne at all times throughout the month long period. Continuing to fly the SP-5B until 1966, VP-48 became the first of the three remaining seaplane squadrons to transition to the P-3B *Orion*. Starting a new career with the P-3 in 1966, the squadron changed home-port to NAS Moffett Field, where they remained until being decommissioned in August 1991.

Prior to their involvement in the South East Asia conflict, the *Marlin* squadrons deployed with their aircraft. This often resulted in long and treacherous flights across large spans of open water, from their respective home-ports, to final destinations somewhere in the Pacific. No longer required to transport individual squadron aircraft from their home-port at NAS North Island, the three remaining squadrons shared the complement of *Marlins* left at NS Sangley Point. Traveling via Navy transports, squadron personnel and material worked their way across the Pacific via Hawaii, Wake Island and Guam, before reaching their final destination in the Philippines. At the onset of each deployment, the incoming squadron would inspect, test fly, and repaint the typical complement of twelve aircraft to reflect their individual designations. Replacement aircraft were stored at the Pacific repair facility located in Japan and were employed when needed to meet fleet commitments.

Rotating in continuous six-month deployment cycles from NAS North Island, the three squadrons flew ASW and ship surveillance missions in the Tonkin Gulf and South China Sea. By comparison to stateside commitments of 500 flight hours or less per month, the squadrons were now flying in excess of 1,100 hours and 170 flights per month, with the same complement of aircraft. Squadron proficiency was now at its peak, and the aging SP-5Bs were pushed to the edge of their operating limits to meet the demanding commitments of operation *Market Time*. Operating from a primary base at NS Sangley Point, they flew tactical patrols in the Philippines area. They supplied detachments of between four and six aircraft to operate with the seaplane tenders *Currituck*, *Salisbury Sound*, and USS *Pine Island* (AV-12) in the forward areas at Cam Ranh Bay, Da Nang and Con Son Islands, South Vietnam.

In the less than placid sea-states of the Pacific, the heavily laden SP-5B *Marlins* often operated near their maximum gross weight limit of 80,000 pounds and required extensive use of JATO to perform their mission under combat conditions. With a crew complement of up to thirteen, added ordnance and a maximum internal fuel load, it became standard procedure to use four, 1,000-pound thrust JATO booster rockets on nearly all mission flights. Whether operating from NS Sangley Point or a remote seadrome somewhere in South Vietnam or Okinawa, the thundering roar of the JATO rocket motors were a common sound.

When operating from NS Sangley Point, it was typical to see local Filipinos in their small *banca* boats, waiting for the seaplanes to jettison the spent rocket motors into the sea. Once dropped, they would attempt to retrieve the empty hulks and sell them back to the Navy for scrap value. However, of the hundreds, or perhaps thousands of JATO bottles used in the Philippines, only a small percentage were actually salvaged and returned.

During tender operations, aircraft moored on buoys were never left unattended by their flight crew. Typically a *buoy watch*, consisting of three flight crew and one flight officer, was established to guard against sabotage or other damage to the aircraft while moored at sea. While on *watch*, personnel were required to maintain routine communication with the tender by use of radio or signal lamp, depending on level of security necessary. Armed and continuously on the alert for enemy swimmers attempting to sabotage the aircraft, buoy crews were ready at a moment's notice to depart the area to a safe zone if instructed by the Air-Boss aboard the tender. This was often the case at Cam Ranh Bay where armed Viet Cong swimmers were common, as was the constant threat of mortar fire from the hills aimed directly at the moored aircraft.

Those not assigned to a *buoy watch* were berthed on the ship which had facilities to sustain and house air-crew personnel. Following a twelve hour patrol, and a long, hot shift on *watch*, crews were eager to go aboard ship for a hot shower prior to their next scheduled flight.

As an additional means to thwart enemy sabotage of the aircraft, tender personnel operated a launch which moved among the moored seaplanes at random intervals, dispensing concussion

A VP-48 SP-5B, BuNo 135542, sits calmly at anchor while its crew goes ashore by launch. (Clay Jansson photo)

The USS *Salisbury Sound* is shown anchored in Cam Ranh Bay with a detachment of *Marlins* moored around her in this 1966 photo.
(U.S. Navy photo)

grenades in the water. Although this acted as an effective deterrent against enemy swimmers, it did little for the tired flight crews attempting to sleep aboard the aircraft.

Battling the broiling tropical sun, and the ever present threat of a typhoon, the flight crews, accustomed to operating as a team, flew eight to twelve hour patrols, monitoring ship traffic with a watchful eye for arms infiltration by sea. Equipped with only a small galley, electric skillet and plenty of imagination, the cook often prepared two or more hot meals aboard the aircraft to feed a typical crew of up to thirteen.

Adapting to the likes and dislikes of their fellow crew members, the cook, typically a voluntary member of the crew who either shared an interest in cooking or just had a veracious appetite, prepared and served each meal between operational commitments. Perishable, fresh food was normally made available for scheduled patrols, whereas back-up or emergency flights were forced to subsist on either boxed or field rations, a less than palatable existence on the long, tiring flights in the tropical climate.

With only the basics to work with, the crews generally gave the cook encouragement and not complaints. When serving thirteen people actively engaged in performing a mission, one had to follow a prescribed procedure. This consisted of serving the non-flying pilot first and working back through the crew positions. A less than rewarding responsibility, one dared not violate this serving order for fear of some form of retaliation from the cook!

There was at least one known incident where an individual had violated this unwritten rule. Somewhere out over the South China Sea, a junior pilot was heard to have rejected his meal, informing the cook that he would let him know when he was ready to eat. Well, as the story goes, when the junior officer finally requested his dinner be served, the cook prepared a special gravy, using a vegetable-based hydraulic fluid. The delayed reaction to this addition to his mashed potatoes, resulted in hours spent in the "head." Although it is doubtful that he realized the reason for his condition, it was rumored that he never questioned a serving time again.

Among the many responsibilities of the seaplane crews, during operation *Market Time*, was the detection and classification of shipping activity in the South China Sea. This process was known as *rigging*. Flying at a dangerously low altitude (between 100 and 200 feet), crews recorded the ship's name, mast, funnel, kingpost configuration sequence, estimated tonnage, course, and speed. When unsure of the purpose of the suspected contact, what seemed to be an easy task, often became a high risk.

Armed with only two door-mounted M-60 machine guns (7.62mm) and wing-mounted five-inch high-velocity rockets, the crews proceeded cautiously when approaching each contact.

In this early 1967 photo, VP-40's Cdr. Longino makes a JATO takeoff from Sangley Point. Barely visible in the water is a local Filipino waiting for the SP-5B to jettison the spent rocket motors. (U.S. Navy photo)

Moored in Cam Ranh Bay in April 1967, VP-40's Crew Five, in BuNo 135479, prepares to spend the night on *watch* in this photo. (U.S. Navy photo)

Ready for an immediate departure at all times, JATO was generally kept installed on the aircraft while on the buoy, as shown here on VP-40's, BuNo 135502 in April 1967. (Author's Collection)

Crews on *buoy watch* during the Vietnam crisis were armed and constantly on the alert for enemy swimmers, attempting to sabotage the moored aircraft, as shown here in 1966. (Author's Collection)

While on *buoy watch*, flight crew personnel are shown rigging shade from an old parachute to help combat the tropical heat.
(U.S. Navy photo)

While *rigging* shipping in the South China Sea in 1966, one VP-40 crew came across this Russian Trawler with sails set—a most unusual sight!
(U.S. Navy photo)

This buoy crew, in VP-40's BuNo 135543, is shown in 1965, receiving hot food delivered from the tender.
(U.S. Navy photo)

Squadrons flying the SP-5B in Vietnam were armed with High Velocity, five-inch rockets, as demonstrated here by VP-31's BuNo 135520.
(U.S. Navy photo)

Hostile small arms fire was not uncommon, and the occasional visual uncovering of concealed deck guns put the crews flying the 78,000 pound, 150 mph seaplane in a highly vulnerable position. An easy target, their only support was from the Air Force early warning communications network (Latch Circuit) which controlled all aircraft in hostile airspace. In the event of an emergency, they would alert the seaplane crew and, if necessary, would provide (if possible) fighter protection against hostile aircraft or ship intervention.

By the time the seaplane squadrons entered the Vietnam conflict, training had reached a pinnacle and the air-crews were a seasoned lot. In order to wear the gold wings of an Aircrewman, the individual was required to demonstrate proficiency in more than one crew position and have a thorough knowledge of the aircraft and its systems. Nonetheless, demonstrated practical knowledge of the aircraft was not the only requirement, for each had to complete rigorous physical tests in survival and undergo psychological fitness, obtaining passing grades in each before being awarded their wings.

As with all naval aviators, survival training was of primary importance, and flying the *Marlin* in the South China Sea and the remote regions of the Philippines was no exception. All flight crew personnel were required to undergo a minimum of one week training, in both land and sea survival techniques, as part of their flight certification. Although ditching at sea was the preferred method for abandoning the seaplane, there was always the distinct possibility of forced survival on land. Individual crew members carried basic survival equipment, including rations, knife, flashlight, flare gun, life vest and a .38 caliber revolver with six rounds of brightly colored tracer ammunition. Use of colored tracer ammunition, not only provided protection, but could be used as a signal if necessary. In addition to personal survival equipment, the aircraft contained two seven-man, MK-7 life rafts, a portable, *Gibson Girl* radio and other primary survival equipment for use on both land and sea.

During the Vietnam conflict, the primary base of operations for the seaplanes was NS Sangley Point located near Cavite City, across the bay from Manila. It was here that routine preventive maintenance was performed, including hourly inspections, minor structural repairs and routine engine changes. Small in physical size, the facilities at NS Sangley Point were old and primitive, affording only the basics for repair. In fact, they were the same buildings and facilities that had been used to support seaplane squadrons for nearly twenty years.

Faced without hanger facilities large enough to accept the seaplanes, support personnel were limited to only a *nose dock* and scaffolding to service the aircraft, with much of the work being performed on the open ramp area. Major structural

On patrol over *Yankee Station*, this cockpit view of an SP-5B shows the pilot's ground track plotter in the center console. Notice the co-pilot's corn cob pipe.
(U.S. Navy photo)

When the SP-5B entered the Vietnam conflict, the crew complement was increased to thirteen. The flight deck accommodated six people, including the Plane Captain, who generally sat on the steps at the entrance to the cockpit, as shown here in 1966. (U.S. Navy photo)

repairs, and high-time inspections, could not be performed locally and were carried out by Shin Meiwa in Japan. Under contract with the Navy since 1960, the facility provided for standard rework, corrosion control, repair and general refurbishment of the *Marlins*.

The archaic repair facilities were not the only outdated accommodations at NS Sangley Point. Personnel stationed at the base were faced with housing which left much to be desired. Squadron flight crew and support personnel were housed in aged, tin Quonset Huts and barracks buildings that, along with their outdated furnishings, had served rotating military personnel since the days of World War Two. Some of these old structures were actually held down by cables to keep them intact during a typhoon. One cannot imagine what it was like to ride out a raging typhoon in these old facilities. With doors and windows battened down, leaving only barely adequate ventilation, personnel were reduced to a bare existence, surviving on only basic rations they provided themselves and a well stocked bar. The deafening roar of the wind and stagnate smell of sweat often lead to days of short tempers in heated solitude—with predictable results.

As squadrons increased in size to fulfill operational commitments, additional personnel were forced to reside in a Barracks Ship (APL) moored at the quay. This appalling derelict of naval history, dating back to the early 1940s, offered less than the basics in comfort, such as defective or nonexistent plumbing, a variety of horrible smells—and those heads! It was not until late 1966 that a new barracks was built to replace these aging facilities.

Due to the nature of their environment, seaplanes were subject to a more frequent preventive maintenance and repair cycle than that of their land-based counterparts. The cost of keeping the Navy's fleet of aging seaplanes free of corrosion was approaching $1 million a year and, by 1966, the $2.5 million *Marlin* was quickly becoming a liability to maintain. Corrosion control on each aircraft required treatments at forty-five and ninety day intervals, often taking up to one week to perform. In addition, a routine fresh water washdown was required after each flight. In forecasting manpower requirements to maintain the predominately saltwater *Marlin*, squadrons had to budget for approximately 32,500 man hours per year to control corrosion.

When assigned to operational exercises, such as tender operations, aircraft were only brought aboard the tender for repairs that either effected watertight integrity or involved major rework, including engine changes. Other routine preventive maintenance and repair was typically performed while the aircraft was in the water.

Shown in 1966 is VP-40's QE-12 in the *nose dock* at NS Sangley Point.
(John R. Gardner photo)

Shown in 1967 is a crewman servicing the starboard A/C generator on an SP-5B moored in Cam Ranh Bay. A trap door in the bottom of the generator compartment provided access to the bomb bay and a place to stand while servicing the electrical components.
(Author's Collection)

Using a series of portable, collapsible scaffolding called *sea-stands*, personnel were able to perform routine service and repairs on the engines and nacelle portion of the seaplane while at sea. Although not foolproof, it was often the only means of making repairs on the water. Attached at the wing root on each side of the engine nacelle, the *sea-stands* were comprised of a series of interlocking walk-ways and support wires that provided a scaffold-like work stand around the engine. Flimsy at best, they had a propensity to collapse at times, dumping personnel, equipment and stands into the sea. While not a common practice, maintenance personnel could effect a complete engine change, using *sea-stands* and other self-contained support equipment while on the water. Other means of servicing the aircraft at sea were limited to accessibility, and often became quite hazardous.

One of the more typical repairs, carried out while the aircraft was on the water, was the replacement of the 40KVA, A/C generators and *Sundstrand* constant speed drive (transmission) units located in the engine nacelles. Accessed through a small service hatch atop each nacelle, these seventy-pound units typically required a two man team for removal. Hoisting the awkward units out through the small opening, while attempting to maintain a solid footing on the oil soaked cowling and wing surfaces, was a challenge for even the experienced and often resulted in disaster, with crewman, tools and equipment falling over the side.

When writing about the successes of the *Marlin* seaplanes, one must also acknowledge the personal commitment of the Martin Aircraft Company to support their aircraft in the field. A policy, dating to the 1940s, Martin provided each squadron with a factory trained field representative who was deployed with the squadron, regardless of location or commitment. These knowledgeable technical representatives, some whom had been with the company since the 1930s, were cross-trained in

VP-50's BuNo 137846 is shown on its way to Shin Meiwa for overhaul in March 1966 on board the USS *Pine Island.* (Shin Meiwa Industries)

every aspect of the *Marlin* and provided invaluable expertise in maintaining the aircraft. Acting as a field liaison between the Navy and the Martin Company, they helped the squadrons resolve the most difficult technical problems under the harshest of conditions. These dedicated service representatives were as much a part of the team as the naval personnel that flew and supported the aircraft, and often they developed a lasting personal relationship with their squadrons.

The flight crews and support personnel that operated the SP-5B *Marlins*, in support of operation *Market Time*, pushed the aging aircraft to their operating limits. Of the thousands of hours flown by these squadrons during this period (1964-1967), few aircraft were lost and none to hostile fire. It would be safe to say that this accomplishment was attributed to the conscientious crews, support personnel and guardian groups associated with the aircraft and their mission. This is not to say that the accomplishments of the seaplane were without sacrifice, for the seaplane squadrons had their share of personal tragedy and loss of life over the years.

The Last Deployment

By the beginning of 1967, only two seaplane squadrons remained in operation, VP-40 and VP-50. In February, VP-40 returned to NS Sangley Point, where they relieved VP-50 for the last time as seaplane squadrons. This hand-off was symbolic in itself, for sixteen years earlier they performed the same ceremony in Iwakuni, as early participants in the Korean War. Leaving their aircraft behind, VP-50 returned to NAS North Island. As with VP-48 the year before, VP-50 proceeded with the transition to the P-3B *Orion* and reassignment to NAS Moffett Field, leaving only one seaplane squadron to defend operation *Market Time.*

Although the men of VP-40 realized with some nostalgia that this would be the last deployment of the *Marlin* seaplane, there was no visible effect on performance or morale, during this tour. Flying lengthy patrols of up to twelve hours, from NS Sangley Point, while supplying detachments to operate with the tenders in Vietnam, the squadron flew an average of over 500 hours per month. Detachments operating in the forward areas with the seaplane tender *Currituck*, flew a rigorous twelve hour on and twelve hour off flight schedule.

As the Navy's last operational seaplane tender, the *Currituck* performed yeoman's service, during this last deployment. Up until 1967, the *Currituck* shared *Market Time* responsibility with its two counterparts, *Salisbury Sound* and *Pine Island.* During the 1966/1967 cycle, the three remaining tenders competed for the Battle "E" award for combat readiness. The *Currituck* walked away with the last pennant, earning the right to wear the big "E" on her bridge and the honor of being the last seaplane tender to receive the coveted award.

Acting as a forward operations seadrome for VP-40, the *Currituck* also served as the flagship for Commander Seventh Fleet, head of operation *Market Time*. By May 1967, the *Currituck* completed its last ten month deployment in support of the Vietnam conflict and returned to home-port in San Diego on May 23, 1967.

Commissioned on June 26, 1944, the *Currituck* was transferred to inactive service, before being decommissioned on October 31, 1967. The *Pine Island* was the first of the three remaining tenders to be retired in 1965, with the *Salisbury Sound* making its last deployment in 1966.

With training proficiency at a peak and morale high, VP-40 continued to excel in all areas of performance. It was during this last deployment that two flight crews achieved the Navy's highest level of flight proficiency in patrol aviation, *Alpha Status*. Awarded for outstanding achievement and personal performance in ASW operations, flight crews of both the commanding officer and executive officer, earned the right to display the red "A" on the bow of their aircraft. While based at NAS North Island in August 1967, Crew Ten, in aircraft BuNo 135533, became the last seaplane crew in the Navy to achieve this respected award. It was perhaps this single accomplishment that led the Navy to choose this particular aircraft for the official last flight in November and eventual restoration for display.

On the evening of March 1, 1967, Crew Ten was scheduled for the nightly patrol from NS Sangley Point. As the crew aboard VP-40's QE-10 performed their pre-flight checks for the twelve hour flight and prepared to launch, the standby crew looked forward to a successful departure which would allow them to stand-down for the balance of the evening. Released from the buoy, the crew reported watertight integrity throughout

VP-40's BuNo 140147 is shown being serviced while towed behind the USS *Currituck* in 1965. This is the same P5M-2 that participated in development projects with VX-1 in 1957.
(U.S. Navy photo)

Seaplanes operating in support of *Market Time* often flew at or near maximum gross weight and routinely required JATO. Shown here is VP-40's 135517 while at Cam Ranh Bay in 1967.
(U.S. Navy photo)

This SP-5B, BuNo 135515, is shown on the stern fueling buoy of the USS *Currituck* in May 1967. (U.S. Navy photo)

Those crews achieving *Alpha* earned the right to display the big red "A" on the bow of their *Marlin*. Shown is VP-40's, QE-7, BuNo 135502 in April 1967. (Author's Collection)

the aircraft and took-up their respective positions for takeoff. Reaching the seadrome, the pilot proceeded with the *mag* (magneto) check. With propellers set at low pitch, and throttles opened to a Manifold Pressure (MAP) equal to that of the field barometric pressure, they proceeded down the seadrome at the equivalent of a high speed taxi.

Check completed, the throttles were reduced, causing the aircraft to settle into the gentle swells of Manila Bay. Proceeding to taxi into takeoff position, the pilot quickly realized a loss of directional control through use of the aircraft's *hydroflaps*. At almost the same instant, the Electrical Panel Operator began to experience an unstable electrical condition, causing circuit breakers to trip off-line and lights to flash.

More concerned about the electrical failure than the loss of steering control, the crew reacted immediately by dispatching someone to check-out the bow and electronic compartments below the flight deck. As the watertight hatch was opened to the electronic compartment, reality overcame shock, for both the bow and electronic bay were flooded and taking on water. Unable to steer the aircraft due to the *hydroflaps* being out of the water, and sinking by the bow, the order was given to abandon ship. As the crew scrambled to break-out the MK-7 survival rafts ahead of the incoming tide, which was driving the sinking seaplane closer to the rocks at the far end of the base, the Radio Operator sent out his first legitimate S.O.S.

At this point, much weighed heavily on the crew's mind, but the biggest topic of discussion was at what depth the PDCs (Practice Depth

Charge), aboard the sinking aircraft, would detonate. Fortunately, the complement of more than twenty-four of the ASW devices did not reach activation depth.

Safely in their life rafts, with help on the way, the crew watched QE-10 slowly disappear beneath the surface of Manila Bay. By morning, only her T-tail and one wing tip was visible above the surf. Shortly thereafter, the aircraft was salvaged and returned to the squadron area. As the SP-5B was brought up the ramp on its beaching gear, the reason for the loss became apparent. Obvious to most, was the open forward port hatch, swinging open and closed with the gentle roll of the tow. Not completely latched, the watertight hatch had sprung open during the *mag* check, filling the two forward compartments with water and leading to the ultimate demise of Navy BuNo 137847. A total loss, the aircraft was stricken from service at NS Sangley Point.

Averaging over one hundred flights per month, during this last *Market Time* deployment, stress and fatigue began to catch up with the tired flight crews and aging aircraft. Between March 19 and April 12, 1967, VP-40 and the *Currituck* participated in the last seaplane-tender operation with the Navy. Operating from Cam Ranh Bay, aircraft patrolled a track known as *Yankee Station* in the South China Sea and Tonkin Gulf, continuously in support of operation *Market Time.* During this period, VP-40 flew a total of 860 flight hours with a complement of seven SP-5B *Marlins,* maintaining an availability of ninety-five percent.

The Final Flights

There were two significant flights of a naval SP-5B seaplane during the Vietnam conflict that climaxed the last deployment of the seaplane with the U.S. Navy. The last combat patrol, originating in a forward area, departed the Cam Ranh Bay seadrome on April 11, 1967. At the control of Cdr. Hugh E. Longino, USN, commanding officer and VAdm. John J. Hyland, USN, Commander Seventh Fleet, VP-40's QE-1 (BuNo 135487) took to the sky for its last twelve hour patrol from a combat zone. This flight marked the closing of the Navy's last seadrome in a forward area and officially ended the last seaplane-tender operation.

In a letter to Cdr. Longino, VAdm. John J. Hyland wrote, "It was a distinct pleasure to participate with you in the final seaplane combat operation in support of *Market Time.* The seadrome/seaplane concept has been with Naval Aviation from its earliest days and has provided flexibility and unstinting support for many operations during its lifetime."

"You can rest assured that you continued to carry-out your surveillance missions faithfully right to the end, and I want to express the appreciation of CINCPACFLT and my appreciation to the person-

Shown with Crew Eleven is Cdr. Hugh Longino, who flew QE-11, BuNo 135543, on the last tactical patrol from NS Sangley Point on May 10, 1967. (U.S. Navy photo via G. Williams)

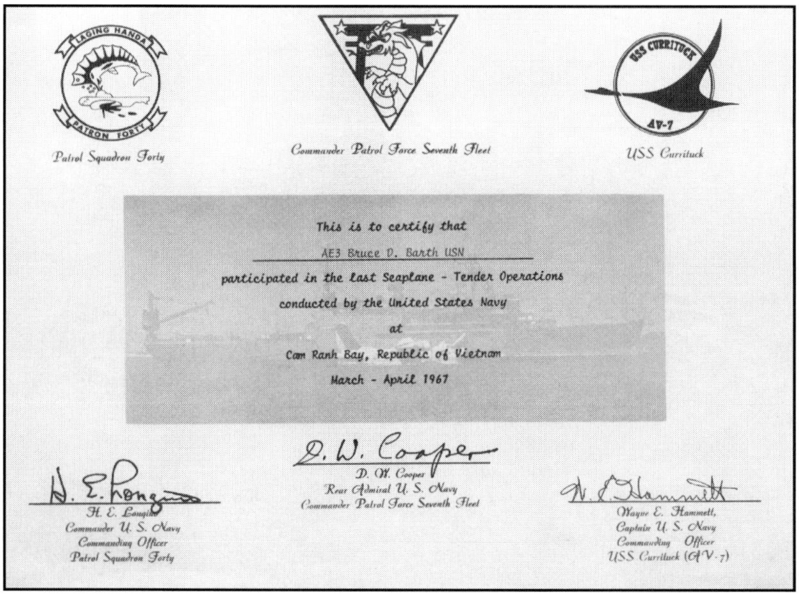
LAGING HANDA
PATRON FORTY
Patrol Squadron Forty

Commander Patrol Force Seventh Fleet

USS CURRITUCK
AV-7
USS Currituck

This is to certify that

AE3 Bruce D. Barth USN

participated in the last Seaplane - Tender Operations

conducted by the United States Navy

at

Cam Ranh Bay, Republic of Vietnam

March - April 1967

D. W. Cooper
Rear Admiral U. S. Navy
Commander Patrol Force Seventh Fleet

H. E. Laughlin
Commander U. S. Navy
Commanding Officer
Patrol Squadron Forty

Wayne E. Hammett,
Captain U. S. Navy
Commanding Officer
USS Currituck (AV-7)

A copy of this certificate was issued to all aircrew and shipboard personnel who participated in the Navy's last seaplane-tender operation at Cam Ranh Bay.
(Author's Collection)

Taken in May 1967, this aerial view of the seaplane ramps at Konan shows several *Marlins* in various stages of being dismantled.
(Author's Collection)

nel of all seaplane tenders and P-5 squadrons, wishing them continued success in their new endeavors."

As noted by Admiral Roy L. Johnson, USN, "As the *Currituck* closes out the seadrome at Cam Ranh Bay, it marks the last combat operation of the seaplane in the U.S. Navy. The tender and seaplane team has performed with distinction in both peace and war, and the completion of this last advance-base operation marks the end of a phase of Naval Aviation dating back to the inception of the naval-air-arm."

"The role played by the tender/seaplane team in the Vietnam War is recognized, as well as its past role in every important naval operation. This team has set a mark of excellence which will always be remembered in the Navy."

In an April 1967 communication, RAdm. Ward recalled, "The *Market Time* patrol, on April 11, with VAdm. Hyland at the controls, marked the end of an era in Naval Air operations."

"Tender supported advanced-seadrome operations have performed many tasks over the years, but never more splendidly than in operation *Market Time* from October 1965 to April 1967. To the officers and men of Patrol Squadron Forty and the

This aircraft, BuNo 135475, was the first P5M-2 accepted by the Navy in 1954 and is shown here in storage at Konan, Japan in January 1967. Acquired by VP-40 one month later, it participated in the last seaplane-tender operation at Cam Ranh Bay in April, before being scrapped in May of '67.
(Shin Meiwa Industries)

By Navy directive, all SP-5Bs operating in the Pacific were to be stricken from service at the Shin Meiwa facility by June 1967. In the background of this June 6, 1967 photo, VP-40's 135543, 135526 and 135475 can be seen with their wings removed.
(Shin Meiwa Industries)

Squadron personnel served as sideboys for this last flight of a Navy SP-5B by VP-40 on November 6, 1967. Crew Ten, in BuNo 135533, also had the distinction of being the last naval seaplane crew to make *Alpha*, as the "A" on the bow indicates.
(U.S. Navy photo)

Piloted by Cdr. Joseph P. Smolinski, VP-40's QE-10 departs the San Diego seadrome on the last official flight by a Navy seaplane squadron on November 6, 1967.
(U.S. Navy photo)

USS *Currituck*, who have the distinction of ending the era, I express my gratitude and congratulations for your outstanding performance of duty and tireless efforts."

- -

For the Navy's last remaining seaplanes in the Pacific, the end was near. In accordance with orders from the Navy Department in early April 1967, all SP-5B *Marlins* operating in the Pacific were to be taken to the repair facility in Japan, where along with those at the facility, would be stricken from naval service.

Commencing on April 18, 1967, VP-40 flight crews began the painfully slow process of ferrying their faithful aircraft to Japan. Dispatched at scheduled release times, the *Marlins* departed NS Sangley Point for Konan, via Iwakuni and Osaka, Japan, before rendezvousing with destiny at the Shin Meiwa facility.

Prior to turning their aircraft over to the Japanese facility for strike, flight crew personnel were responsible for removing all classified material and ordnance typically carried aboard; however, the cavernous SP-5B, having been used extensively over the past six-months in support of the Vietnam conflict, held many obscure hiding places—something was bound to be overlooked.

On one such flight, BuNo 135502 took its turn in the queue for the final flight into history. Touching down in the busy Konan Harbor on May 2, the aircraft taxied towards the waiting launch that would guide them to the buoy. Searching for a piece of misplaced personal equipment, one of the crew pulled up the internal bilge decking to retrieve it. He found more than he expected. For there, sitting in the bottom of the bilge, was one of the *Marlin's* primary pieces of defensive armament—a complete M-60 machine gun! How this could have happened was not questioned, it was what to do with it that posed a problem. As the plane taxied towards the launch, the nervous crew made a decision to throw it overboard. What a personal prize it would have made, but one not worth the consequences, should one get caught attempting to smuggle it home through U.S. Customs in Hawaii.

On May 10, 1967, Crew Eleven departed NS Sangley Point on the squadron's last tactical patrol in support of operation *Market Time*. By this point, many of the aircraft had already departed the Philippines for their eminent demise in Japan. The commanding officers number one aircraft (QE-1) had been one of the first to leave. Of the few remaining aircraft available, QE-11, BuNo 135543, was chosen for the last patrol. Flown by Cdr. Longino and Capt. Herbert S. Ainsworth, Commander Fleet Air Wing Eight, the ten hour flight officially ended the last deployment of the Martin SP-5B *Marlin* with the U.S. Navy. The closing of the seadrome at NS Sangley Point brought to a close nearly thirty years of seaplane operations in the Philippines.

The twelve SP-5B aircraft assigned to VP-40, and those undergoing rework or stored at the Shin Meiwa facility, comprised the total complement of *Marlins* in the Pacific. By direction of the Navy, these were dismantled and scrapped at the Japanese facility. None returned to the United States, and all were destroyed. Working within operational commitments, the process of turning in their aircraft was completed by May 12, and the squadron prepared for the long return trip home to NAS North Island, via the familiar Navy C-118B transports provided by VR-21.

From their first deployment in support of the Vietnam conflict in February 1965, through their final deployment in May 1967, the SP-5Bs of VP-40 flew a total of 2,331 flights, and 14,534 hours, in support of operation *Market Time*. Because this flight data represents only one of the three squadrons that participated in this effort, one can easily expect these figures to more than triple, when considering the accomplishments of VP-48 and VP-50 during this period.

The seaplane had long since outlived its usefulness in the eyes of the military; however, when called to task one last time, this single accomplishment by the Navy's last seaplanes would become a significant part of aviation history.

Following the squadron's return to home-port in June, they resumed routine operational commitments in the San Diego area with the Navy's last remaining SP-5B seaplanes. Awaiting final notice to transition to the P-3 *Orion*, the squadron proceeded to reduce their inflated complement of personnel, necessary for *Market Time* operations and to schedule training in the new aircraft.

By October, it was announced that the squadron would change home-port to NAS Moffett Field and proceed with the transition to the land-based Lockheed P-3B by late November. This transition completed the phasing out of the seaplane from the naval inventory.

The last official flight, by a naval seaplane squadron, was scheduled for November 6, 1967. In a formal ceremony at NAS North Island, attended by hundreds of military and civilian personnel and press, squadron members formed one hundred yards of sideboys, as the last crew marched stoically to their aircraft. As the band played, VP-40's QE-10A (BuNo 135533) took to the air for one final flight around the San Diego area.

Piloted by the last commanding officer of a seaplane squadron, Cdr. Joseph P. Smolinski, Jr., with RAdm. C. A. Karaberis, Commander Fleet Air Pacific, the *Marlin* completed its flight without incident, officially ending the career of the seaplane with the Navy. As the big SP-5B returned to the squadron area following the flight, taps filled the air in tribute to the fifty-six years of seaplane activity at NAS North Island. All remaining seaplanes at NAS North Island were taken to a remote part of the base known as the *bone yard*, where, stripped of all useful parts, they awaited final destruction.

FOR OFFICIAL USE ONLY — PATROL SQUADRON FORTY — FOR OFFICIAL USE ONLY

C/O FLEET POST OFFICE
SAN FRANCISCO, CALIFORNIA

FLIGHT SCHEDULE FOR MONDAY, 6 NOVEMBER 1967

CDO: LCDR JARWIN
SDO: LTJG GARNETT

PHONE: 5982
HIGH TIDE: 0112/1124
LOW TIDE: 0454/1918

SUNRISE: 0610
SUNSET: 1655
DUTY SECTION: 3
DUTY CREW: 5

FLT NO	PREFLT LAUNCH T/O BUOY ETE	MODEX TYPE BUNO CREW	PILOT IN COMMAND CO-PILOT NAV TACCO	MISSION	FUEL (GAL) (HRS)	CLEARANCE VOICECALL	ORDNANCE RATIONS BRIEF REMARKS
1.	1000 1100 1130 1200 0+30	QE-10 SP-5B 135533 CREW TEN	CDR SMOLINSKI CDR SUROVIK LCDR HENSON LCDR SMITH LCDR WILLIAMS	SPECIAL + RADM KARABERIS	1800 6+00	DD-175 QE-10	CHARLIE NOTE 5

1. ARTICLE 201.c OF THE NORIS AIR OPERATIONS MANUAL HAS BEEN COMPLIED WITH.
2. ALL COMMUNICATIONS IAW FAW-14 OPORD 201-66, ANNEX F.
3. THE FOLLOWING OFFICERS AND ENLISTED MEN ARE SCHEDULED FOR LOW PRESSURE AND NIGHT VISION ON MONDAY, 6 NOVEMBER 1967, AT 0745, AT NAS MIRAMAR.

SUMNER, ATR3	DWYER, ATRAN	DANIELS, AE2	SITZMANN, AME3
SURINA, ATN3	HILES, AMSC	BARTH, AE3	PARHAM, ADJ3
NEWMAN, ATN3	MAGSAMEN, ADRC	KING, AO3	WOOD, AE2

   HEALTH RECORDS WILL BE REQUIRED AND WILL BE AVAILABLE IN THE DUTY OFFICE MONDAY MORNING. TRANSPORTATION WILL LEAVE THE SQUADRON AREA AT 0630.
4. THE AROUND THE CLOCK EFFORT OF NAS NORTH ISLAND IN SUPPORTING THIS SQUADRON'S SEAPLANE OPERATIONS IS SINCERELY APPRECIATED. TO THE CRASH BOAT CREWS WHO WAITED WATCHFULLY AS OUR AIRCRAFT DEPARTED AND RETURNED; TO THE PUBLIC WORKS DEPARTMENT PERSONNEL WHO WERE CONTINUOUSLY AT WORK ON THE SOUTH BAY SEADROME LIGHTS; TO THE GCA CONTROLLERS WHO GUIDED US TO SAFE LANDINGS IN BAD WEATHER; AND TO THE SUPPLY, AMD, O & R, AND SALVAGE PERSONNEL WHO COMBINED TO KEEP THE AIRCRAFT PARTS AVAILABLE TO SUPPORT OUR MAINTENANCE EFFORT; AND TO ALL OTHERS WHO CONTRIBUTED THEIR TIME AND EFFORT IN OUR BEHALF, A SINCERE WELL DONE.
5. EVENT ONE REPRESENTS THE LAST SCHEDULED FLIGHT OF A SEAPLANE OPERATED BY THE UNITED STATES NAVY. IT IS APPROPRIATE THAT THE FINAL CHAPTER IN THE SEAPLANE'S LONG HISTORY IS WRITTEN BY A SQUADRON LONG ASSOCIATED WITH TYPE AIRCRAFT AND WITH A HIGH REPUTATION FOR PROFESSIONALISM. THOSE OF US THAT KNOW AND UNDERSTAND THE SEAPLANE VIEW HER PASSAGE INTO HISTORY WITH A CERTAIN AMOUNT OF NOSTALGIA. ON THE OCCASION OF HER RETIREMENT, LET US TREAT THE SEAPLANE WITH THE RESPECT, ADMIRATION AND DIGNITY BEFITTING HER MANY YEARS OF LONG FAITHFUL SERVICE.

SUBMITTED BY:

J. D. HENSON, LCDR, USN
OPERATIONS OFFICER

APPROVED BY:

J. P. SMOLINSKI, JR, CDR, USN
COMMANDING OFFICER

MABUHAY OLD FRIEND

DEPARTMENT OF THE NAVY
UNITED STATES PACIFIC FLEET
AIR FORCE
PATROL SQUADRON FORTY
FPO SAN FRANCISCO 96601

FF12/VP40/10:rjf
1650
Ser:

From: Commanding Officer, Patrol Squadron FORTY
To: The Men of Patrol Squadron FORTY
Subj: Letter of Commendation

1. The seaplane commenced its active duty with the United States Navy in the second decade of the century. It established the major share of world's records accumulated by the Navy in the early years of aviation. In World War II, the PBY *Catalina* gained prominence island hopping across the Pacific to be fol lowed by the PBM *Mariner* during the Korean conflict. In the Vietnam crisis, the P5M *Marlin* achieved its first taste of advanced base combat operations. The story of the seaplane in the U.S. Navy has been a proud one, and now you have had a part in insuring that this tradition has survived through its entire span.

2. During the period of 19 March through 12 April 1967 Patrol Squadron FORTY shared with USS *Currituck* the privilege of conducting at Cam Ranh Bay Vietnam, what will probably be the last seaplane-tender operation conducted by the United States Navy. During this period an aircraft was maintained on station continuously in support of Market Time anti-infiltration efforts accumulating a total of 860 flight hours. This was done with seven SP-5B aircraft at the tender maintaining an availability of 94.38 percent.

3. You are commended for your part in insuring that historians will speak highly of the seaplane from start to finish. The final phases have been accomplished under trying conditions, with old equipment, shortages, and extended deployments. The professionalism you have demonstrated, the long hours you and your shipmates have expended, and above all, the pride of accomplishment you have exhibited, have all worked together to make this possible. The trust placed in you by history has been well met. Well done.

4. A copy of this letter will be made a part of your permanent Service Record.

HUGH E. LONGINO

Copy to:
BUPERS

The Martin P5M
Marlin

Chapter Eight

The Last Seaplane Squadron

Patrol Squadron Forty was first designated on September 1, 1948; however, the roots of this squadron date back to the Navy's first Martin PBM *Mariner* seaplane squadron, VP-55. This was the first of four different designations assigned the squadron between 1940 and 1950. Following re-commissioning in January 1951, the squadron continued their legacy as a highly competitive seaplane squadron, seeing action in both the Korean War and the Vietnam conflict. Operating continuously in this capacity until November 1967, VP-40 played the last role in the history of the seaplane with the U.S. Navy.

With the commissioning of VP-55, at NAS Norfolk, in August 1940, the Navy took delivery of their first PBM *Mariner* seaplane, the PBM-1. Early production of the *Mariner* was limited, with VP-55 receiving the first twelve of the twenty produced.

In 1941, the squadron designation was changed to VP-74. Throughout the war years, they proved themselves a threat to German U-boats, achieving six confirmed kills. In the tense period that followed the war, the squadron was re-designated Medium Seaplane Squadron Ten (VP-MS-10), while based at Naval Station Coco Solo, Panama, in November 1946. Adjusting to peacetime activities, such as flying air-sea rescue and guarding the Panama Canal Zone, it wasn't until September 1948 that the squadron was assigned their final designation, VP-40.

Continuing with their operational commitments, VP-40 remained at NS Coco Solo for the next sixteen months. Ordered to prepare for decommissioning as early as January 1950, VP-40 commenced a transfer of all squadron aircraft to NAS Norfolk on January 9. The process was completed four days later, on January 13. On January 25, 1950, with an appropriate ceremony, one of the oldest naval seaplane squadrons (VP-40) was decommissioned at NAS Norfolk. This brought to a close ten years of flying the Martin PBM *Mariner,* in both combat and peacetime, throughout the Atlantic, Caribbean and the Southeast Pacific.

With the outbreak of the Korean War, the Navy once again recognized the need to utilize the versatility of the seaplane as part of their defensive arm. During the post-war period, many seaplane squadrons were decommissioned or placed on inactive reserve status. Within months of initial hostilities, the Navy embarked on a program to strengthen its existing complement of (six) seaplane squadrons (VP-34, 42, 45, 46, 47, 49). Three reserve squadrons were activated in late 1950, VP-661, 731 and 892, two of which (731 and 892) departed for the Korean theater shortly there after. Two additional squadrons were re-formed in January 1951, VP-40 at NAS North Island and

This 1951 photo shows VP-40's first Crew One with PBM-5, CA-1. Shown kneeling, second from left is Cdr. Vadym V. Utgoff. Notice the JATO bottles attached to the port waist hatch. (Vadym Utgoff Collection)

Shown here is an early VP-40 PBM-5, CA-4, BuNo 84787, at Oppama, Japan in 1951.
(Robert Orne photo)

VP-40's CA-8 is shown here being launched at Iwakuni, Japan in 1951. An RAAF *Sunderland* can be seen in the background.
(Vadym Utgoff Collection)

VP-44 (January 29) at NAS Norfolk. This brought the total number of squadrons flying the PBM *Mariner* to eleven.

Patrol Squadron Forty was re-formed on January 20 under the direction of Cdr. Vadym V. Utgoff, USN, an accomplished veteran in seaplanes during World War Two, with VP-34's *Black Cats*. The newly formed squadron was assigned the Martin PBM-5 and PBM-5S *Mariner* aircraft, the latest models in the PBM series.

Following a brief period of formation and training, VP-40 departed home-port on May 26, 1951 for the first of two deployments to the Western Pacific, in support of the Korean War. Leaving the security of NAS North Island behind, the flight of nine, heavily laden, (maximum gross weight of 62,000 pounds) *Mariners* ventured out over the Pacific on the first leg of their arduous journey to Iwakuni, Japan.

Flying in sections of three aircraft, their route across the Pacific took them via Ford Island, Hawaii; Johnston Island, Kwajalein Atoll, Guam and Oppama, Japan before reaching their final destination some nine days later. Operating under command of Fleet Air Wing Six, the squadron arrived at the RAAF Base in Iwakuni where they relieved VP-892. Commencing on June 9, VP-40 began its first six-month period patrolling the Tsushima Straits, flying anti-submarine cover for replenishment groups in the Yellow Sea, Sea of Japan, and the Philippines, along with weather reconnaissance flights for other fleet forces.

This P5M-1, QE-12, BuNo 130303, is shown in 1957. VP-40 was the first West Coast squadron to receive the P5M-1 and the first to deploy the *Marlin* under operational commitments.
(U.S. Navy photo)

1951 Squadron Insignia

1955 Squadron Insignia

1959 Squadron Insignia

While based at NS Sangley Point in 1959, VP-40 changed their squadron insignia and adopted the motto *Laging Handa*. This local ceremony included the Mayor of Cavite City, the Governor, and Cdr. Bill Cockroft, C.O., left.
(Bill Cockroft Collection)

Located on the island of Honshu, the Australian air-station at Iwakuni was the primary base of operation during this first deployment. Sharing responsibilities with crews flying the RAAF *Sunderland,* VP-40 flew hundreds of missions over the Korean waters, in all kinds of weather, mostly at night. Flying lengthy patrols of ten to fourteen hours each, crews were often faced with weather so rough that it was difficult to maintain altitude within 200-300 feet and stay within ten degrees of course. During this deployment the squadron was forced to evacuate their aircraft three times to avoid approaching typhoons. Aircraft were typically flown to one of three alternate seadromes in the area; Buckner Bay, Okinawa; Yokosuka, Japan; or NS Sangley Point, Philippines. Despite the foul weather, long hours and maintenance problems, VP-40 was able to meet all operational commitments and achieve a high degree of availability with their aircraft throughout this deployment.

Completing their first deployment on December 10, 1951, VP-40 was relieved by another NAS North Island based *Mariner* squadron, VP-42. Departing Iwakuni, the squadron's nine aircraft retraced their route across the Pacific for the fifty-eight hour return flight to home-port, arriving in San Diego on December 17. At a formal awards ceremony on May 17, 1952, one-hundred seventeen Air Medals were presented to VP-40 aircrews who participated in the Korean campaign. VP-40 remained at NAS North Island, performing routine operational commitments and training until August 20, 1952 when they commenced their second deployment. During this period at home-port the squadrons complement of PBM-5 aircraft was increased from nine to twelve.

On September 2, 1952, VP-40 arrived at NS Sangley Point where they again relieved VP-892. Operating from such places as Iwakuni and Okinawa in Japan, Boko Ko in the Pescadores and Sangley Point in the Philippines, they performed ASW exercises, ship surveillance, weather reconnaissance, air-sea rescue and tender operations. While under command of Fleet Air Wing Two, they patrolled the Formosa Straits until March 28, 1953. Completing their second deployment, VP-40 returned to the United States.

For their achievements during these early deployments, squadron personnel were awarded the China Service Medal, Korean Presidential Unit Citation, Korean Service Medal and the United Nations Service Medal.

Following VP-40's return to NAS North Island in 1953, the squadron transitioned to the Navy's newest medium-range seaplane, the Martin P5M-1 *Marlin,* on April 30. Actively engaged in war-time responsibilities, VP-40 had yet to adopt a squadron namesake. Recognizing a need to acknowledge their latest acquisition of the *Marlin* in some way, squadron personnel held a contest to choose a new design for the squadron insignia and an appropriate slogan. They even contacted Walt Disney Studio, who for years had offered the services of their art department to create a squadron insignia. Although Disney came through with a proposed design, the winning entry came from within the squadron. Combining the name of the new aircraft with that of their role in ASW, the *Fighting Marlins* were born.

Introduced to the fleet in 1952, the P5M had yet to be tested under actual fleet conditions. In July 1953, VP-40 became the first squadron to conduct training exercises with the new aircraft and the seaplane tender, *Salisbury Sound.* As these were the first tests of their kind with the P5M, exercises were documented for use in training future shipboard and squadron personnel in the handling of the new aircraft.

Patrol Squadron Forty was the first squadron to deploy the P5M under actual operational commitments, in January 1954. Despite their unfamiliarity with the new aircraft, the squadron averaged eight planes in commission and was able to meet all operational commitments, during this deployment to NS Sangley Point.

As the Korean War drew to a close, VP-40 began a series of peacetime deployments to the Western Pacific area through 1958. Rotating in six-month deployment cycles with other seaplane squadrons, they performed a myriad of operations, including ASW exercises, weather reconnaissance, air-sea rescue and tender operations. These deployment cycles included January to July 1954, April to September 1955, June 1956 to January 1957 and April to September 1958.

In October 1955, the squadron was presented with their first Battle Efficiency "E" Award. Awarded to squadrons displaying the highest overall proficiency rating for operational readiness, hours flown and flight safety record, this award was the first to be presented by the Navy since being temporarily suspended during the Korean War.

In August 1959, VP-40 made a permanent change of home-port from San Diego to NS Sangley Point. It was during this deployment in November 1959 that VP-40 changed their insignia and adopted the motto, "LAGING HANDA," meaning *Always Ready* in Tagalog, the native language of the Philippines. While based at NS Sangley Point, the squadron initiated a transition to replace their existing P5M-1 *Marlins* with the improved P5M-2 model. Commencing on October 4, 1960, they completed the transition by July 24, 1961.

For over four years the squadron flew patrols in the Philippines and South China Sea area. Battling the ever present typhoon and tropical heat, they took part in numerous air-sea rescue operations and honed their ASW skills with the new P5M-2. During this period, the squadron worked closely with surface and air forces, both Chinese Nationalist and American, in support of the Quemoy and Matsu crisis. Working with the seaplane tenders at Boko Ko in the Pescadores, the crews from VP-40 routinely patrolled the Formosa Straits, earning the Armed Forces Expeditionary Medal for their achievements.

By late 1963, the squadron was ordered to resume a routine six-month rotational deployment schedule. Relieved by a detachment of *Marlins* from VP-48 based in Iwakuni, VP-40 returned to NAS North Island as their home-port. Arriving in March 1964, the squadron was presented with their second Battle "E" Award for the 1962-1964 competitive cycle.

During the Vietnam conflict, VP-40's *Fighting Marlins* flew their SP-5B seaplanes in support of operation *Market Time* in the Tonkin Gulf and South China Sea. Rotating continuous deployments to NS Sangley Point from NAS North Island, they operated routine patrols in the Pacific between 1965 and 1967.

In February 1965, VP-40 returned to NS Sangley Point and started patrolling the familiar waters of Southeast Asia. Flying tactical patrols from NS Sangley Point, with detachments operating in the forward areas from the seaplane tenders, squadron personnel adjusted to the rigorous routine of defending the Tonkin Gulf from communist shipping activity. Relieved by VP-50 on August 25, 1965, VP-40 completed their deployment cycle and returned to home-port. In a ceremony at NAS North Island on December 10, 1965, eighty-

While permanently based in the Philippines, VP-40 made the transition to the P5M-2 in 1961. This SP-5B, QE-1, BuNo 135487, is shown on routine patrol over the South China Sea in 1967. It was in this aircraft that Cdr. Longino flew the last combat patrol by a Navy SP-5B. (U.S. Navy photo)

Shown here in 1967 is the Navy's last Crew One to fly the Martin SP-5B. L/R kneeling; Weatherly, AX3; Gardner, AMH3; Smith, AO3; Potter, AX2; Plaugher, AX2; Barth, AE3; Renoud, AX3. L/R standing; Vallett, ADR1; Ltjg. Sorrick, Lt. Vincent, Cdr. Smolinski, Ltjg. Boling, Ltjg. Buie.
(U.S. Navy photo)

Shown at NS Sangley Point in 1967 is VP-40's QE-7, BuNo 135502. This was the Executive Officer's aircraft during the last deployment of the Martin SP-5B. Both Crew Seven and Crew One made *Alpha* during the last deployment.
(John R. Gardner photo)

1967 Special Edition Insignia

After 26 years of flying the P-3 *Orion,* VP-40 still retains the QE tail code and squadron insignia from the seaplane era, as shown in this 1992 photo.
(Nick Veronico photo)

Shown with two squadron P-3C Update III *Orions,* VP-40 personnel form the *Fighting Marlins* insignia in this 1992 photo taken at NAS Moffett Field. (U.S. Navy photo)

seven crew members were awarded the Air Medal for their first operational deployment in support of the Vietnam conflict. (A similar ceremony was held at NAS North Island in 1952, following VP-40's first deployment in support of the Korean War.)

Following extensive training exercises at NAS North Island, VP-40 returned to NS Sangley Point again in February 1966. Rejoining the tenders off the coast of South Vietnam, they continued their support of operation *Market Time.* Working closely with Vietnamese and U.S. surface units, the squadron did its share to prevent the infiltration of communist arms and munitions into South Vietnam by sea. On September 1, 1966, having completed their deployment cycle, VP-40 was relieved by VP-50 and returned to NAS North Island.

In February 1967, VP-40 relieved VP-50 at NS Sangley Point for the last time. As the Navy's last seaplane squadron, VP-40 once again took-up station with the tender, *Currituck,* in the forward area at Cam Ranh Bay. Between March and April of 1967, they completed the last seaplane-tender operation with the U.S. Navy. Accepting the fact that they had completed their last deployment as a seaplane squadron, VP-40 returned to NAS North Island in June.

In a formal ceremony at NAS North Island on November 6, 1967, VP-40 officially retired the naval seaplane, giving way to the new technology and the latest addition to the patrol ASW fleet, the Lockheed P-3 *Orion.* By the end of November the squadron commenced transition to the P-3B and changed home-port to NAS Moffett Field, where they remained for the next twenty-six years. In December 1993, the squadron changed home-port to NAS Whidbey Island.

Throughout the squadron's long and tenuous career in flying boats, it is believed that only four aircraft were lost due to operational accidents. On June 15, 1951, a PBM-5S was lost on an instrument climb-out from Iwakuni during adverse weather conditions. During a typhoon evacuation in June 1960, a P5M-1 was lost in a landing accident at Iwakuni. In August 1962, a P5M-2 was lost in the mountains of the Philippines, due to navigational error, while on an ASW exercise, and in March 1967, an SP-5B was lost, due to a component structural failure, resulting in an accidental sinking off NS Sangley Point.

Squadron Insignia

In the sixteen years that VP-40 flew as an operational seaplane squadron, several designs were chosen to represent their squadron insignia. A painting was commissioned by the author in 1989 which portrays the history of these designs, during the many years VP-40 flew as an accomplished seaplane squadron. It was a time when the squadron was so proud of its past, they refused to give up the *Fighting Marlins* reputation and insignia, even after the transition to the P-3 *Orion.* From the "Bee" of 1951, that symbolized their mission strength to the *Laging Handa* motto adopted in 1959, the *Fighting Marlins* remain a significant part of naval aviation history. After twenty-six years and 180,000 accident-free flight hours flying the P-3 *Orion,* the squadron still retains their insignia and namesake from the seaplane era.

Design Credit

1951: First squadron patch designed by Petty Officer Robert Maley. Submitted by Commanding Officer Cdr. Vadym V. Utgoff.

1955: Designed by Chief Petty Officer Richard Claydon to commemorate their new P5M-1 *Marlin* aircraft. Submitted by Commanding Officer Cdr. Winton C. Sharpe.

1959: Third design by Commanding Officer Cdr. Irving G. Cockroft. Same basic design and colors used through 1993.

1967: Designed and submitted by squadron personnel under the direction of Commanding Officer Cdr. Joseph P. Smolinski, Jr. Produced as a final acknowledgment to an **"End Of An Era"** in naval seaplane aviation.

This painting was commissioned by Bruce D. Barth in 1989 and is on display at VP-40's home-port in Whidbey Island, Washington.

Patrol Squadron Forty

Commanding Officers

1948 through **1950**

September 1, 1948 to December 4, 1948	Cdr.	H. F. Perrenot
1948 to 1949	Cdr.	Thomas E. L. McCabe
1949 to January 25, 1950	LCdr.	William D. Harrington

1951 through **1967**

January 20, 1951 to 1952	Cdr.	Vadym V. Utgoff
1952 to 1953	Cdr.	Miles S. Whitener
1953 to 1954	Cdr.	Joseph M. Kellam
1954 to 1955	Cdr.	Winton C. Sharpe
1955 to 1957	Cdr.	Donald C. Coy
1957 to 1958	Cdr.	Jack W. Clinton
1958 to 1959	Cdr.	Clifford B. Curtis, Jr.
1959 to 1960	Cdr.	Irving G. Cockroft
1960 to 1961	Cdr.	E. W. Myers
1961 to 1962	Cdr.	Jack S. Laney
1962 to 1962	Cdr.	Norbert P. Vegelahn
1962 to 1963	Cdr.	John R. Johnson
1963 to 1964	Cdr.	H. Ken Cooley
1964 to 1965	Cdr.	Frank J. Schneider
1965 to 1966	Cdr.	Harry J. Hinden
1966 to 1967	Cdr.	Hugh E. Longino, Jr.
1967 to 1968	Cdr.	Joseph P. Smolinski, Jr.

Service Awards

China Service Medal	1953
Korean Presidential Unit Citation	1953
Korean Service Medal	1953
United Nations Service Medal	1953
Battle Efficiency “E” Award	1955
Armed Forces Expeditionary Medal	1959
Battle Efficiency “E” Award	1964
Vietnam Service Medal	1965
Republic of Vietnam Campaign Medal	1965
Republic of Vietnam Meritorious Unit Citation	1967
Navy Unit Commendation	1967

Tail Designator

1948 - 1957: “CA”   1958 - 1967: “QE”

The Martin P5M *Marlin*

Chapter Nine

One More Flight

In early 1968, the Smithsonian Institution proposed the idea of preserving one of the Navy's SP-5B *Marlins* for future display. Acting on the authority of the Naval Air Systems Command, Washington, D.C., the Naval Air Rework Facility (NARF) at NAS North Island was assigned the project of removing VP-40's QE-10 (BuNo 135533), from the salvage area and returning it to operational flight status for one last flight.

Under the direction of Supervising Project Planner and Estimator, Richard H. Benbough, a comprehensive plan was put into effect to restore the SP-5B to air-worthiness. Rather than restore the aircraft for static display at NAS North Island, or transport it via ground to another facility, the Navy elected to fly it to NAS Patuxent River, where it was to be donated to the Smithsonian Institution's proposed National Armed Forces Museum in Washington.

In the short time that had elapsed since the aircraft's historic flight on November 6, 1967, it lay derelict in the salvage yard, stripped of its most usable components, including the engines. The undertaking of finding certifiable parts to restore the aircraft for flight, and locating divers knowledgeable in rigging the underwater cables for the beaching ramps, were only a few of the challenges facing the project. Once replacement R-3350-32WA engines were located and installed, the aircraft was towed to the NARF area where the restoration work would be accomplished. Stored outside, with no intention of being returned to service, the *Marlin* was not treated with a protective coating, and therefore also required treatment for corrosion control and painting prior to use.

Perhaps the biggest challenge of the project was locating a flight crew with current qualifications to fly the aircraft. The fleet training squadron, VP-31, had stopped training flight personnel for the SP-5B in early 1967, and the crews from the last squadron to fly the aircraft had been dispersed within the fleet, having lost their proficiency in the *Marlin.* The responsibility for assembling a qualified crew to fly the aircraft across country was assigned to Aircraft Ferry Squadron Thirty-Two (VRF-32), NAS North Island.

Navy pilots are required to undergo routine proficiency checks to maintain their flight status in aircraft type. These checks are typically performed by senior instructors, who are qualified in these aircraft and are able to certify another pilot's proficiency in the aircraft type. Since the Martin SP-5B was phased out of the naval inventory the

VP-40's QE-10 is shown being removed from storage prior to restoration in early 1968.
(U.S. Navy photo via Richard H. Benbough)

VP-40's BuNo 135533 is shown here undergoing restoration at the Naval Air Rework Facility, NAS North Island in early 1968. (U.S. Navy photo via Richard H. Benbough)

Shown in June 1968, with BuNo 135533, is the Navy's last flight crew to fly the Martin SP-5B. (U.S. Navy photo via Richard H. Benbough)

previous year, the availability of a qualified instructor to perform a proficiency check was not available. As it turned out, the last qualified Patrol Plane Commander (PPC) in the Navy, with a current proficiency rating in the SP-5B, was LCdr. Donald R. McGarrigle, attached to VRF-32. This finite qualification became the controlling factor in the completion of the aircraft's restoration and subsequent flight across country.

Working towards this objective, Benbough expedited the restoration of the aircraft, and the rerigging of the launching ramps, at NAS North Island which had been dismantled in late 1967. Following completion of the restoration, the aircraft required pre-flight and flight tests, prior to the actual flight, to ensure both aircraft and crew proficiency. By May, the final crew had been identified and assigned to the project and a target date of July 8, 1968 was established for departure. Flight testing commenced in June without incident, and the project continued on schedule.

In addition to the operational complement of flight crew members, several prominent people also accompanied the *Marlin* on its final journey across country. The supplementary crew included Capt. Jack Sullivan, USNR, representative for the Martin-Marietta Corporation, and two members of the National Armed Forces Museum Advisory Board; Col. R. Calland, USMC (Ret.), and Maj. John Elliott, USMC (Ret.). This special flight was further highlighted by the addition of WAVE, Ltjg. Marianne Blackburn, assigned as Assistant Public Affairs Officer for VRF-32.

The project was completed on schedule, and on July 6, 1968 a formal dedication ceremony was held at NAS North Island to mark its completion. During the ceremony, Capt. W. Shockey, com-

BuNo 135533 is shown here being brought up the ramp at NAS North Island in early July 1968. This photo was taken just days before the last flight on July 8 and clearly shows the special nose art on the bow. (U.S. Navy photo via Richard H. Benbough)

This photo shows SP-5B, BuNo 135533, departing the San Diego seadrome for the last time on July 8, 1968. This last departure ended nearly fifty-seven years of naval seaplane activity in the San Diego area. (U.S. Navy photo via Richard H. Benbough)

manding officer NARF, NAS North Island formally presented Capt. Hugh Tate, commanding officer, VRF-32, with the logs for the aircraft. Present at the ceremony was Capt. J. Sullivan who, as a representative for the manufacturer, Martin-Marietta Corporation, presented Capt. Tate and his crew with a model of the aircraft. Guest speaker for the event was RAdm. C. Karaberis, Commander Fleet Air, San Diego, who dedicated the historic aircraft to the young people of the United States. Completed in record time, given the circumstances, the project was a success, and the aircraft was readied for departure from NAS North Island as scheduled.

For the last time, on Monday, July 8, 1968, Navy SP-5B, BuNo 135533 thundered down the San Diego seadrome to take its place in history. At a gross weight of 68,000 pounds, the *Marlin* lifted graciously off the water and headed East over the city that knew it so well. With a destination some 3,000 miles away, the crew of thirteen realized that this last flight of a naval seaplane would propel them into Naval Aviation history. Scheduled to arrive at their final destination of NAS Patuxent River, on the morning of July 12, the *Marlin* made two scheduled stops enroute. The first being NAS Corpus Christi, a seaplane base since the 1940s, and the second, NAS Jacksonville, Florida, also an ex-seaplane base that could accommodate and service the large flying boat.

Flying conditions were favorable, and the big *Marlin* experienced no major problems on its long trek across country—attributed to the thoroughness of the refurbishment personnel at the NARF facility. On the morning of July 12, 1968, the aircraft left Jacksonville on the last leg of its journey. Arriving at NAS Patuxent River just before noon, the aircraft circled the seadrome, allowing the crew to savor the few remaining minutes in the air before touching down in history, for this was the *last time the Marlin would ever fly.*

QE-10's Arrival at NAS Patuxent River

The ceremony awaiting the arrival of the *Marlin* rivaled that of its last flight with VP-40 some nine months earlier. Attended by civilian and military dignitaries alike and press from around the country, this ceremony marked the closing of a significant chapter in Naval Aviation. Making a long, slow approach to the seadrome at NAS Patuxent River, QE-10A made a picture perfect landing on the calm water. Flanked by both military and civilian craft, the big seaplane taxied towards the awaiting crowd. LCdr. McGarrigle navigated the aircraft into position on the buoy while the beach crew readied the ramps to retrieve the *Marlin* for the last time. When the throttles of the powerful Curtiss-Wright R-3350 radial engines closed for the last time, and the aircraft was brought up the inclined ramp, it was perhaps LCdr. McGarrigle who realized most the significance of this flight. For he alone would become a part of aviation history, as the last pilot to fly the *Martin P5M,* the Navy's last production seaplane.

Capt. Huge Tate, officer in tactical command of the last flight, departed the aircraft by launch, transporting the log books for the Navy's last seaplane ashore. In a formal display of tradition, Capt. Tate presented VAdm. Thomas F. Connolly, Deputy Chief of Naval Operations for Air, with the aircraft's log books. In turn, VAdm. Connolly presented the books to John Nicholas Brown, Smithsonian Regent and chairman of the Institution's National Armed Forces Advisory Board. Mr. Brown, having had a long association with the Navy, served as Assistant Secretary of the Navy for Air from 1946 to 1949. Also present at the ceremony was John Rauth, then President of Martin-Marietta Aerospace Corporation.

A special nose art was applied to the Navy's last seaplane for its final flight to NAS Patuxent River in July 1968. Represented were the official Navy crew members aboard the historic flight and those from Aircraft Ferry Squadron Thirty-Two. In addition to placement on both sides of the nose, the AIRFERRON THREE-TWO flash itself was displayed on the center of both sides of the vertical stabilizer fin. The red "A," denoting the *Alpha Status* of the last operational crew to fly the aircraft, was moved from the nose to the antenna housing on the rear of the tail.

Accepted and delivered to the Navy on May 17, 1956, BuNo 135533 was assigned to service with VX-1 and FASRON-102 until January 20, 1958,

On the morning of July 12, LCdr. McGarrigle is shown taxiing the Navy's last *Marlin* towards the awaiting ceremony at NAS Patuxent River. (U.S. Navy photo)

Shown on the buoy at Patuxent River, BuNo 135533 prepares to be brought up the ramp. In the foreground, notice the main beaching gear sitting at the ready. (U.S. Navy photo)

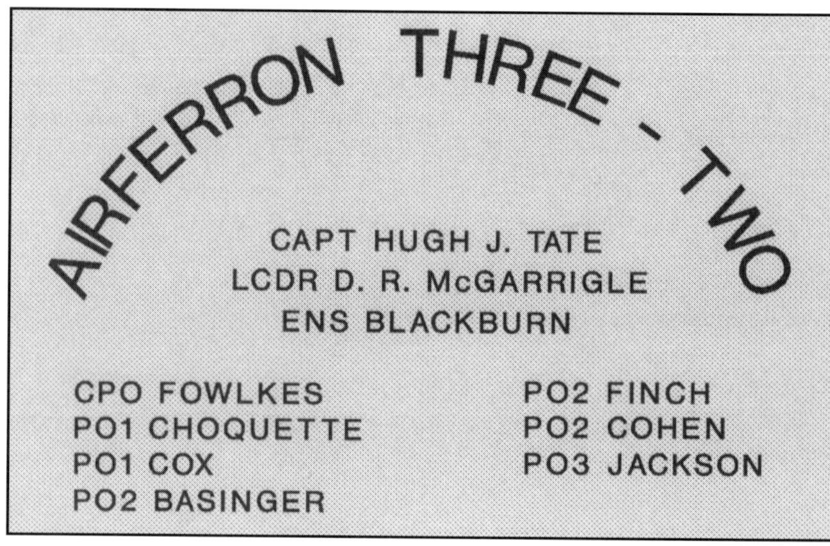

This special nose art, acknowledging the last official Navy crew members to fly the SP-5B across country, was applied to the bow of the aircraft for the flight. (Author's Collection)

VP-40's QE-10 is shown here on the ramp at NAS Patuxent River on July 12, 1968. (U.S. Navy photo)

John Nicholas Brown, center, is shown officially receiving the aircraft's log books on behalf of the Smithsonian Institution. (U.S. Navy photo)

when it was transferred to its first fleet squadron, VP-44. There, only a few months, it was transferred to O & R (Overhaul and Repair) at Norfolk and on to Martin in Baltimore for modification updates. Following ASW modifications, it was assigned to VP-56 where it remained until June 8, 1961. It was then transferred to BUWEPS in Norfolk and then to O & R for overhaul. On October 11, 1963, it was transferred to VP-48 based at NAS North Island. Less than one year later it was assigned to VP-40, on March 18, 1964.

The aircraft remained at NAS North Island during the Vietnam conflict and was routinely transferred between the Navy's three remaining seaplane squadrons based in San Diego. Starting with February 10, 1965, it was transferred to VP-48. Then on September 30, 1965, it was transferred to O & R (NAS North Island) for rework prior to departing for St. Augustine, Florida, where a lengthy overhaul was performed by the Fairchild Corporation. Returned to NAS North Island on April 26, 1967, the aircraft was temporarily assigned to VP-50 before its final assignment to VP-40 on June 29, 1967. Following the squadrons change of home-port to NAS Moffett Field on November 16, 1967, the aircraft was transferred to the storage facility at NAS North Island.

Of the almost 300 P5M *Marlins* produced between 1951 and 1960, this lone surviving example now stood silently in the Washington sun. During his acceptance speech in July 1968, Mr. Brown noted with sorrow, that this truly was an "End Of An Era" in Naval Aviation.

Storage at NAS Patuxent River

Sealed and preserved in an environmentally safe coating, the restored aircraft was placed in storage, solemnly awaiting its turn to be placed on display at the proposed National Armed Forces Museum. By the early 1970s the aircraft had not yet found a home for permanent display, as the National Armed Forces Museum had not been completed as planned. Although sealed and protected, the aircraft was fighting a losing battle against time. Stored outside at a remote portion of the base, it was slowly becoming a victim of the elements and losing the once, proud significance it had earned in 1968.

At about this time, the National Museum of Naval Aviation, located in Pensacola, Florida, was growing and in need of aircraft for display. The SP-5B in storage at NAS Patuxent River was soon to find a home. In the mid 1970s, plans were completed to include the SP-5B as part of the Museum's inventory of display aircraft. Mounted on its original beaching gear, with wings removed, the *Marlin* was transported by barge to its final

BuNo 135533 is shown here in storage at NAS Patuxent River in April 1971.
(Joseph Handelman photo)

destination. Traveling from NAS Patuxent River, via NAS Norfolk, the aircraft arrived at Chevalier Field, Pensacola, where it was reassembled and moved to the Museum grounds.

Pensacola Naval Air Station

Officially opened on June 8, 1963, the National Museum of Naval Aviation, located on the sprawling grounds of NAS Pensacola, offers an excellent collection of original aircraft, representing both fighters and patrol aircraft from throughout the Navy's long association with aviation. As the Museum increased in size, the Martin SP-5B *Marlin* took its place beside a growing number of historic seaplanes including the Consolidated PB2Y *Coronado* and PBY *Catalina;* Grumman HU-16 *Albatross,* JRF *Goose,* J4F-1 *Widgeon* and J2F-6 *Duck;* and the original 1919 Curtis NC-4 flying boat. In its collection, the museum also contains a replica of the Curtis A-1 *Triad,* an amphibious version of the first seaplane built in 1910.

Although not totally representative of all naval seaplanes over the years, this fine collection of vintage aircraft *is* representative of the greater majority of which that have managed to withstand the ravishes of time. And, many are the last remaining examples of their type.

In this 1975 photo, BuNo 135533 is shown being transported by barge from NAS Patuxent River to the National Museum of Naval Aviation in Florida.
(U.S. Navy photo)

1992

Completely restored to original condition prior to its last flight in 1968, the aircraft is painted its original Insignia White and Light Gull Gray colors dating to the Vietnam era and stands proudly just outside the main Museum building. Working with limited funding to restore the many aircraft on display, the outward appearance of the Museum's *Marlin* seems to have withstood the Florida sun well over the years; however, it is lacking the distinctive nose art that once identified its historic final crew members, and the big red "A" on the rear antenna housing that denoted the *Alpha Status* of the last operational crew to fly the aircraft, VP-40's Crew Ten.

Of the thousands of visitors that attend the Museum annually, it can be said with great certainty that many will remember the Martin *Marlin.* For this great aircraft has the distinction of being one of four remaining Martin flying boats in existence today and, is the only surviving example of the *Marlin* series. It was also this particular aircraft, BuNo 135533, that ended the reign of the naval seaplane, an era spanning 57 years, 1911 to 1968.

Still painted its original Insignia White over Light Gull Gray colors from the Vietnam era, BuNo 135533 is shown here in 1993 without the nose art from its historic flight in 1968.
(Herb Montgomery photo)

Shown in 1993, VP-40's QE-10 sits just outside the main building at the National Museum of Naval Aviation.
(Herb Montgomery photo)

The Martin P5M
Marlin

Chapter Ten

Conclusion

In all, nearly 1,700 medium-range seaplanes of the PBM and P5M models were manufactured by the Martin Aircraft Company between 1939 and 1960. The PBM class consisted of three basic models; PBM-1, PBM-3 and PBM-5. Updated to the latest -S and -S2 configuration, the PBM-5 continued to operate with VP-50 until July 31, 1956 when they were officially retired from fleet service.

In the late 1940s, a limited number of PBM-5s were manufactured in a amphibious configuration. Designated the PBM-5A, the project was terminated after only forty were built, due in part to the excessive weight of the retractable landing gear conversion. The only example of the PBM *Mariner* series to have avoided destruction is BuNo 122071, a PBM-5A. Delivered to the Navy on May 6, 1948, it was first assigned to VP-33. Following its useful service life with the Navy, the PBM was sold on the surplus market for civilian application and sat derelict for many years in Southern California before being serviced for a one-time ferry flight to Arizona. Owned by the Smithsonian Institution's National Air and Space Museum, this aircraft is on display at the Pima Air Museum located in Tucson, Arizona where it is presently undergoing restoration.

There are only two other known examples of the PBM series, however both are under water wrecks that have yet to be recovered. Located on the bottom of Lake Washington, lies the remains of two PBM-5 *Mariners* that crashed in the same year while on ferry flights to the nearby Boeing facility. On February 20, 1949, Navy BuNo 59254 crashed and sank during a hard landing on the South side of the lake. Later that year, on May 6, BuNo 59172 sank near the Boeing seaplane ramp, following an accident while taxiing. Apparently, both were *high-time* aircraft on their way to storage at the Boeing facility at the time the accidents occurred. There were no fatalities on either crash and both aircraft were stricken from naval service.

The P5M was produced in only two basic models, the dash-one and dash-two. The last squadron to fly the *Marlin* and retire the series in November of 1967 was VP-40 at NAS North Island. The only remaining example of this series is an SP-5B, BuNo 135533 owned by the National Museum of Naval Aviation, Pensacola, Florida. The Museum's example was manufactured in 1956 and first delivered to VX-1. It last saw service with VP-40 at NAS North Island in November 1967. Restored by the Navy in 1968, it was flown to NAS Patuxent River and later transferred to the Museum for display.

Unlike the PBM, there are no known salvageable wrecks of the P5M in the United States. Those that are known to have crashed, while on deployment outside the continental U.S., saw either extensive damage or deep water submersion and are not considered salvageable at this time.

This PBM-5 shown at NAS Norfolk in 1946, BuNo 59349, was the single prototype used for the forty PBM-5A amphibians built through 1949. (Warren M. Bodie Collection)

Shown at the Pima Air Museum in Tucson, Arizona in 1993 is PBM-5A, BuNo 122071. This aircraft is the only surviving example in the PBM *Mariner* series.
(Nick Veronico photo)

Shown here at the National Museum of Naval Aviation in Pensacola, Florida is SP-5B, BuNo 135533. Of the 283 *Marlins* produced, this aircraft is the only surviving example.
(Author's Collection)

Perhaps the reason that more of these huge military seaplanes did not survive lies simply with the fact that they were just that—military seaplanes. Designed primarily as a military electronics platform for a role in ASW, the non-pressurized fuselage of the P5M was never configured to accommodate seating or cargo.

Unlike amphibious type seaplanes that could be flown to a central storage facility, the demise of the obsolete P5Ms came swiftly. Considered a heavy aircraft at 78,000 pounds, the *Marlin* required a complex beaching system for launch and retrieval from land. The aircraft itself required the use of three, large and cumbersome, floating, beaching gear assemblies (wheels). The inclined ramp required a complex system of underwater cabling and ground crew personnel to operate the equipment. Without facilities to retrieve the P5M, the only alternative was to leave it in the water which, for its size, would require large docking facilities and extensive preventive maintenance to control corrosion. Without a civilian application, ramping or storage facilities, there was no apparent market for the aircraft. As a result, their fate was sealed where their careers ended.

The predominant history of the Navy since 1911 was not based solely on that of their surface naval fleet, but a blend of air and sea. The seaplanes played a major role in nearly all aspects of naval aviation during their fifty-six year career. As a formidable opponent to geography, the seaplane had virtually no bounds. It could operate from remote seadrome waterways anywhere in the world under the harshest of conditions, be a formidable threat to submarine warfare, and save the lives of countless downed flyers in rescue efforts.

Nevertheless, as the foremost military superpower in the world, the United States decided to relinquish its interest in the seaplane in 1967 when dissolving the Navy's last seaplane squadron, VP-40. And yet, many nations today still retain the seaplane as a strategic military aircraft, and those that do, have taken on some form of Martin design from the 1940s through the 1950s. Perhaps we should ask ourselves—*why didn't we?*

Shown at Kaneohe, Hawaii in 1945 is VPB-32's PBM-5, *Loose Goose*, BuNo 84648. The PBM was the backbone of the Navy's flying boat fleet during World War II and served in every theater of the War.
(Harold Stetson Collection)

The Russian Beriev A-40, is a multipurpose, twin-engine, jet-powered amphibian capable of over 800 km/hr which is currently in use in Russia.
(Author's Collection)

The Martin P5M
Marlin

Chapter Eleven

Related Art Work and Squadron Humor

Throughout the history of naval aviation, squadrons routinely adorned their aircraft with various forms of nose art or other means of personal identification. Typically depicting their mission or individual personalities, these works of art personified the crews that flew them. Their application and use was at the discretion of the squadron's commanding officer and was not generally governed by a naval directive. Of the eleven U. S. squadrons that flew the P5M, it is believed that only three used some form of personal art work on their aircraft, VP-40, VP-45, VP-48. Of these three, VP-48 was the most predominate. During their involvement in the Vietnam conflict, VP-48 adorned many of their aircraft with various forms of nose art, including one P5M-2 (135482) with a *shark mouth* painted on the bow. In addition, the squadron used a *three-color chevron* design on the tail of each aircraft. The use of nose art by VP-40 was limited to one known example in 1960 of a P5M-1 (135470) with the words, *Looney Gooney*, painted below the cockpit. Patrol Squadron Forty-Five decorated their *Marlins* with a single *chevron* on the bow above the aircraft number.

Given such endearing names as *Pig-Boat*, *Pelican* and *Penguin*, the P5M was also the subject of many squadron artists, whose creative talent and imagination seemed limitless when it came to the *Marlin*. Their creations took on many forms and were often displayed in books published by individual squadrons following deployment. This chapter contains a compendium of art work, advertising and related humor, by numerous artists associated with Martin flying boats throughout the 1940s, '50s, and '60s. They are a reflection of the many talented individuals who were attached to the squadrons or associated with the Martin Aircraft Company during the period in which they were manufactured.

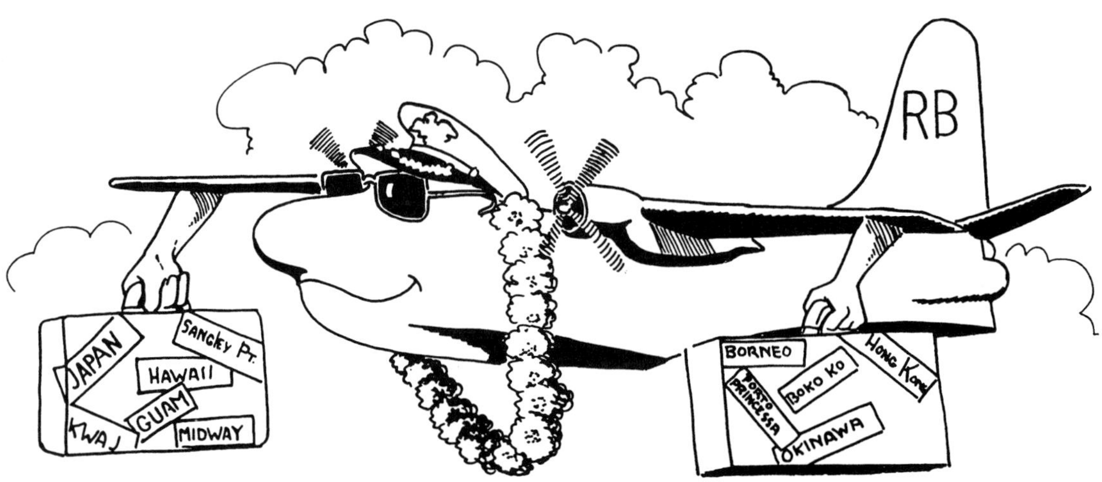

P - Boat Pilots Prayer

Late one night in a South China Bay,
a young P-Boat pilot knelt down to pray.
The moon shown down on the buoy that night,
and the time had come to make things right.
I guess I'll have a word with the man, he said,
just a short talk before I rest my head.
Some guy up there in the black cold space,
must hear me speak from this foul smelling place.
Sir, I've been to many an exotic and bright spot,
like this, where years past, the Spaniards fought.
I've seen the world in its many shades,
and somehow its splendor never fades.

But tonight I've got some things to say,
about the way a P-Boat pilot earns his pay.
Like this watch, for one, I have to stand,
couldn't we lock'em and leave'em like's done in New Zealand?
Here we sit, getting crusted with salt,
when I'd rather taste the hops and malt.
Yes, how about a nice ten hour flight,
flown at the same speed as a good sized kite?

And crashing through the chop and briny swell,
try that at night for a special hell.
Impossible tasks are placed in our way,
so soon in the mirror you detect some gray.
Take a *SAR* and a *Barrier* the Admiral will say,
but my god, in one plane, on the same day?
Sleep an hour and be ready to go again,
and in between, keep the crew out of sin.

So please old man somewhere in the sky,
let me see the day when I'll have to fly,
only twenty hours a month and then
plenty of time off with a feminine friend.

And sure enough a voice was heard,
saying "copied all charlie most of your word."
Five-by-five he heard the voice say,
"say again all after, — I kneel down to pray."

By Ltjg. David E. Lang Patron Forty, 1960

RNZAF THE SOCIETY FOR USN

PRESERVATION OF SEA PLANE PERSONNEL

in SOUTH the PACIFIC

Be it hereby known that he whose name doth appear above has manifestly demonstrated the proper amount of scorn for all those who land on straight and clear paths and smooth concrete runways and in witness thereof, is entitled to lifetime membership in this society.

He is entitled to make high speed taxi and open sea landings, and is also cleared to launch and land in all liquid media, be it clear or murky.......................................

This certificate was drawn by VP-40 artist, Dave Lang, for SEATO exercise *Sealion* with the RNZAF in 1960. (via Bill Cockroft)

This 1955 painting by Martin artist, Keith Bagley, was widely published and was often seen hanging in squadron spaces as late as 1967. (via "Mac" MacVicar)

WELL, CHECK THE FUEL PANEL ANYWAY, WE CHECKED EVERYTHING ELSE!

LET'S RUN THROUGH TH' ABORT PROCEDURE AGAIN, JIM.

Combat Aircrewmen Must...

Do any Crewman's job in flight . . .

Be able to perform First Aid . . .

Understand electricity . . .

Be intelligent and athletic . . .

Send and receive CW 12 words a minute . . .

Know the elements of survival

NO TRESPASSING!

To range far and fast over the oceans in the tireless watch of our far-flung coasts . . . to summon the might of our battle fleets before trouble gets within striking distance . . . that's a job for Martin PBM-1's . . . swift and potent "air cruisers" that extend the "reach" of national defense. Mighty, twenty-ton Patrol Bombers . . . packed with fire power for defense and striking power for attack . . . Martin PBM-1's have been joining the Navy in ever-increasing numbers that put new teeth in "No Trespassing."

THE GLENN L. MARTIN COMPANY, BALTIMORE, MD., U. S. A.

Martin
AIRCRAFT

Builders of Dependable Aircraft Since 1909

TRADE MARK

1934—THE CHINA CLIPPER, *first of the big Martin flying boats to fly the Pacific, was the first over-ocean luxury liner, confirmed the practicality of regularly scheduled over-ocean transport flights.*

1937—THE MARINER LINE, *begun in this year, had no peers in World War II as patrol bombers and search planes. And hundreds of Martin Mariners, including PBM-5 shown above, still serve the Navy.*

1948—THE CAROLINE MARS, *world's record-holder for total passengers carried in a single flight, is today's giant descendant of the original "Old Lady" Mars, largest flying boat in the world when it was built in 1942.*

RUGGED NEWCOMER
to a Select Circle

WITH its strikingly longer afterbody hull and husky construction, the new Martin XP5M-1 is designed to operate from any of the Earth's open waters 95% of the year! This rugged patrol craft makes safer landings with less pitching and bouncing. Tail surfaces are protected from waves by added buoyancy in the stern. Take-off time and distance are shortened. And the hull is designed to eliminate "skipping off" in landing.

Ideal for patrol work over rough seas, this husky twin-engine plane is expected to have greater range than any other flying boat in its class when completely developed. It's the latest in a great family of Martin flying boats! THE GLENN L. MARTIN COMPANY, BALTIMORE 3, MARYLAND.

Martin
AIRCRAFT

Builders of Dependable Aircraft Since 1909

Individual flight crew patches were always unique and told a story. A reserve squadron based in Seattle, VP-892 was activated in 1950. Comprised of members with a variety of individual talents, Crew Eight deemed themselves *a butcher, a baker and a candlestick maker*, as shown in this 1951 crew patch. (Neil McLeod Collection)

The Martin P5M
Marlin

Appendices

Model 162
PBM "Mariner" Data

"X" Series Development Aircraft

Designation: **X** = Experimental **P** = Patrol **B** = Bomber **M** = Martin

Model 162A: Serial Number X19168

XPBM-1: BuNo 0796 First prototype development aircraft produced in PBM series. ①

XPBM-2: BuNo 1247
Second prototype PBM. Increased fuel capacity from 2,700 gallons to 4,815 gallons, structural re-enforcement and modifications for catapult launching. Although tested and evaluated successfully in 1942 at the Naval Air Material Command, Philadelphia, Pennsylvania, this model was not put into production. It was stricken from naval service on June 30, 1944.

XPBM-3E: BuNo 6456
Second production PBM-3. Radar development aircraft for AN/APS-15 search radar.

XPBM-5: BuNo 45275 and BuNo 45276
Two prototypes were used to evaluate uprated engines. Modifications included the use of Pratt & Whitney R-2800-34 Twin Wasp engines rated at 2,100 horsepower and Hamilton Standard three-bladed aluminum propellers.

XPBM-5A: BuNo 59349
Single prototype. Modified PBM-5 with retractable tricycle undercarriage.

Production Models and BuNo Data
1937 to 1949

PBM-1: 1246, 1248 - 1266
PBM-3R: 6455, 6457 - 6504
PBM-3C: 6505 - 6747, 01650 - 01673
PBM-3S: 01674 - 01728, 48125 - 48163
PBM-3D: 45205 - 45274, 45277 - 45404, 48124, 48164 - 48223
PBM-5: 45405 - 45444, 59000 - 59348, 84590 - 84789, 85136 - 85160, 98602 - 98616
PBM-5A: 122067 - 122086, 122468 - 122471, 122602 - 122613 ⑥

Note:

① Contract Date: 06-30-37 First Flight: 02-18-39 Last PBM Delivery: March 1949
② Production models modified for special application used suffix designators;
B = Twenty-seven PBM-3Cs were delivered to Great Britain under a lend-lease program, 1943/44
C = Increased defensive armament, improved crew protection, uprated engines (-3)
D = PBM-3C, BuNo 6656 served as prototype. Uprated engines and additional armament (-3)
E = Special purpose radar and electronics (-5)
G = USCG application. ASW electronics and armament deleted (-3S, -5G, -5A)
M = PBM-5E, BuNo 98606. First modified as a -5N all-weather. Later converted for missile tests
R = Transport version. Strengthened floor, passenger seating, no armament (-3)
S = Special purpose radar, ASW detection equipment, different armament (-3, -5)
③ Total production in PBM series, 1937 to 1949: 1,368 including "X" models
④ Last fleet squadron to operate: VP-50, Alameda, Ca. 1956
⑤ Last training squadron to operate: ATU-501, Corpus Christi, Tx. 1958
⑥ Four additional PBM-5 models were converted to PBM-5A. BuNo 59253 - 59256

Model 162
PBM "Mariner" Data

PBM Squadron Development By Year and Model

Year/ Model	Fall 1940	Dec 1941	Jul 1942	Jan 1943	Jul 1943	Jan 1944	Jul 1944	Jan 1945	Jul 1945
PBM	2	1	1	1	1	5	6	3	-
PBM-3	-	-	-	3	2	2	2	1	-
PBM-3C	-	-	-	7	5	4	3	1	1
PBM-3D	-	-	-	-	-	3	12	11	4
PBM-3S	-	-	-	-	5	5	6	9	1
PBM-5	-	-	-	-	-	-	-	6	13
PBM-5E	-	-	-	-	-	-	-	-	2

PBM Statistics By Model

Model:	PBM-1	PBM-3C	PBM-3R	PBM-3S	PBM-3D	PBM-5	PBM-5A
Engine Type:	R2600-6	R2600-12	R2600-12	R2600-12	R2600-22	R2800-22	R2800-34
Manufacturer:	Wright	Wright	Wright	Wright	Wright	P & W	P & W
Horsepower:	1,600	1,700	1,700	1,700	1,900	2,100	2,100
Gross Weight:	48,000 lbs	50,664 lbs	55,000 lbs	53,225 lbs	56,853 lbs	58,000 lbs	60,300 lbs
Range:	3,375 miles	2,300 miles	2,980 miles	3,130 miles	2,085 miles	2,350 miles	1,920 miles
Max Ordnance:	6,000 lbs	8,000 lbs	8,000 lbs	8,000 lbs	8,000 lbs	8,000 lbs	8,000 lbs
Max Speed:	205 mph	205 mph	207 mph	207 mph	190 mph	182 mph	183 mph
Service Ceiling:	19,200 feet	17,700 feet	17,700 feet	13,000 feet	15,900 feet	21,000 feet	19,100 feet

Post-War Reserve PBM Squadrons

Squadron:	VP-661	VP-731	VP-892
Reserve Station:	Anacostia, D.C.	Grosse Isle, MI.	Seattle, WA.
Tail Code:	A	I	T
Date Activated:	09-15-50	10-03-50	08-04-50
Redesignated:	VP-56, 03-02-53	VP-48, 02-04-53	VP-50, 02-04-53

Appendix C

Model 237
P5M "Marlin" Data

Production Models and BuNo Data
1946 to 1960

Designation: **P** = Patrol **5** = 5th Model **M** = Martin

• **XP5M-1:**	Single Prototype. Converted PBM-5, BuNo: 98616 Redesignated M-270 in 1951	
• **P5M-1:**	Total Production = 167 ① ⑤	
	BuNo:	124910 - 124914, 126490 - 126511, 127696 - 127719 130265 - 130351, 135452 - 135473
• **P5M-1G:**		1284 - 87, 1295 - 97
• **P5M-1T:**		149825 - 149831 ② ③
• **P5M-2:**	Total Production = 116 ① ⑤	
	BuNo:	135474 - 135543, 137846 - 137848, 140140 - 140150 141252 - 141258, 147926 - 147937
• (France)		146440 - 146445, 147539 - 147542 ⑦
• **P5M-2G:**		1312, 1318 - 1320
• - to USN:		149832 - 149835 ③

Significant Dates

• **XP5M-1:** First Official Flight:	-	05-30-48	Baltimore, Md.
• **P5M-1:**			
First Official Navy Flight:	-	06-21-51	Baltimore, Md.
First Atlantic Squadron:	VP-44	04-23-52	Norfolk, Va.
First Pacific Squadron:	VP-40	04-30-53	North Island, Ca.
First Deployment:	VP-40	WestPac	February to July 1954
• **P5M-2:**			
First Official Flight:	(BuNo 130286)	08-53	Baltimore, Md.
First Production Flight:	-	04-29-54	Baltimore, Md.
First Navy Delivery:	(BuNo 135475)	06-23-54	Norfolk, Va.
First Squadron Delivery:	VP-47	10-18-54	Alameda, Ca.
Last Production Delivery:	-	12-20-60	U.S.N.
Last Squadron Flight:	VP-40	11-06-67	North Island, Ca. ⑥
Last Official Flight:	VRF-32	07-08-68	North Island, Ca.

Note:

① Official production figures vary. Information based on most accurate data available.
② USCG P5M-1G: Redesignated P5M-1T (training) when returned to USN in 1960
③ USCG P5M-2G: Assigned USN BuNo when returned in 1960. Same for ②
④ 09-30-58: Tail code designator system changed for naval aircraft
⑤ 09-18-62: Uniform aircraft designator system established. Changed to SP-5A and SP-5B
⑥ Last AirLant P5M-2 Squadron, VP-45. 12-30-63
⑦ Ten P5M-2s were delivered to France under a lend-lease program. No suffix designator change

Appendix D

Performance and Dimensional Data Martin PBM and P5M

Model	PBM-5	XP5M-1	P5M-1	P5M-2
Wingspan: ⑤	118 feet	118 feet	118 feet	118 feet
Length: ⑤	79 feet 10 inches	90 feet 10 inches	90 feet 3 inches	92 feet 8 inches
Height:	28 feet 10 inches	36 feet 7 inches	35 feet 2 inches	32 feet 8 inches
Width:	10 feet	10 feet	10 feet	10 feet
Gross Weight: ⑩	58,000 pounds	60,000 pounds	73,000 pounds	78,000 pounds
Engines: 2 ea.	R-2800-34	R-3350-26W	R-3350-30WA ⑧	R-3350-32WA
Rating:	2,100 H.P.	2,700 H.P.	3,250 H.P.	3,450 H.P.
Manufacturer:	Pratt & Whitney	Wright Cyclone	Wright Compound	Wright Compound
Propeller dia.	14 feet 8 inches	14 feet 8 inches	15 feet	15 feet
Max. Speed: ① ③	238 mph	249 mph	269 mph	276 mph
Cruise:	147 mph	150 mph	157 mph	159 mph
Range:	2,800 miles	1,400 miles	3,000 miles	3,000 miles
Service Ceiling:	21,000 feet	21,000 feet	21,000 feet	21,000 feet
Crew: ⑦	7 to 13	7 to 10	7 to 10	7 to 13
Armament:	8 x .50 cal. guns	6 x 20mm guns	2 x 20mm guns ②	2 x 20mm guns ②
Fuel: ④ ⑨	3,488 gallons	3,488 gallons	3,975 gallons	3,993 gallons

Note:

① Airspeed conversion factor:
Nautical (knots) to Statute (mph): x 1.15078
Statute (mph) to Nautical (knots): x 0.8689

② Turret removed on later models with "S" modification

③ Performance data reflects base model configuration. Actual performance varied according to configuration and operating conditions

④ Figures are maximum and include two auxiliary 575 gallon drop tanks where applicable

⑤ Dimensions are typical and do not reflect wing-tip lights, protruding electronic components and or armament on tail assembly

⑥ Draft in sea water at gross weight of 78,000 pounds (64 pounds per cubic foot at step);
With beaching gear: 96.5 inches Without beaching gear: 58.5 inches
Note: Draft greater in fresh water than sea water

⑦ Actual crew complement determined by ASW and ordnance configuration

⑧ Several versions of the Curtiss-Wright R-3350 engine were used on the P5M-1 through 1960 including -30W, -30WA, -32WA, -36W, -36WA

⑨ Fuel Grade: Standard = 115/145 Octane Alternate = 100/130 Octane
Fuel Density: 6 pounds per U.S. Gallon

⑩ Figures represent manufacturers normal gross weight specification. Maximum allowable gross weight overload approximately 10%

Appendix E

Martin P5M Publications

The following represents a basic listing of technical publications published by the Martin Aircraft Company on the P5M series seaplanes. Dates shown reflect a general publication date and could be preceded by an earlier version or replaced by a later revision. Regardless of the date of issue, the publication code remained the same for a given title.

Publication Code	Title	Release Date
AN 01-35EJA-1	Flight Handbook, Navy Model P5M-1	March 1, 1952
CO 01-35EJA-1A	Supplemental Flight Handbook, Navy Model P5M-1 Confidential Operating Data	March 1, 1952
AN 01-35EJA-2	Handbook Maintenance Instructions, Navy Model P5M-1	June 15, 1954
AN 01-35EJ-3	Handbook Structural Repair, Navy Model P5M-1	March 15, 1952
AN 01-35EJA-4	Illustrated Parts Breakdown, Navy Model P5M-1	September 15, 1954
CO 01-35EJ-501	Supplemental Operation, Maintenance and Repair with Parts Breakdown, Navy Model P5M-1. Confidential	March 15, 1955
AN 01-35EJA-6	Handbook Inspection Requirements, Navy Model P5M-1	December 15, 1952
AN 01-35EJB-1	Flight Handbook, Navy Model P5M-2	June 15, 1957
AN 01-35EJB-1A	Supplemental Flight Handbook, Navy Model P5M-2	September 15, 1954
AN 01-35EJB-2	Handbook Maintenance Instructions, Navy Model P5M-2	December 15, 1956
AN 01-35EJB-3	Handbook Structural Repair, Navy Model P5M-2	June 15, 1954
AN 01-35EJB-4	Illustrated Parts Breakdown, Navy Model P5M-2	September 15, 1955
AN 01-35EJB-502	Supplemental Operation, Maintenance and Repair with Parts Breakdown, Navy Model P5M-2	July 15, 1955
AN 01-35EJB-6	Handbook Inspection Requirements, Navy Model P5M-2	May 1, 1954
NAVAER 01-35EJB-513	Handbook of Operation, Maintenance, Repair with Parts Breakdown. Special Stores Version, P5M-1 and P5M-2	November 15, 1957
NAVWEPS 01-35EJA-7	Inspection Requirements, Intermediate and Major Navy Model P5M-1 and P5M-2	August 1, 1963

Additional publications were available from the manufacturer relative to specific systems and components in the model series, including training manuals. NATOPS (Naval Aviation Training and Operating Procedures Standardization Program) Manuals on the P5M series were published in five revisions by the Navy Department between 1960 and 1967.

Last Significant Flights
Martin SP-5B

April 11, 1967	**May 10, 1967**	**November 6, 1967**	**July 8, 1968**
Last Combat Patrol	**Last Tactical Patrol**	**Last Operational Flight**	**Final Flight**
Squadron: VP-40 Crew 1	Squadron: VP-40 Crew 11	Squadron: VP-40 Crew 10	Squadron: VRF-32 Special
Location: Cam Ranh Bay, SVN	Location: NS Sangley Point	Location: NAS North Island	Location: NAS North Island
Aircraft: BuNo: 135487	Aircraft: BuNo: 135543	Aircraft: BuNo: 135533	Aircraft: BuNo: 135533

(Crew Listing)

Cdr.	H. Longino	①	Cdr.	H. Longino	①	Cdr.	J. Smolinski	①	Capt.	H. Tate ②
VAdm.	J. Hyland	③	Capt.	H. Ainsworth	④	RAdm.	C. Karaberis	⑤	LCdr.	D. McGarrigle
Lt.	W. Able		Lt.	W. Able		Cdr.	G. Surovik	⑥	Ltjg.	M. Blackburn
Lt.	P. Schneider		Ltjg.	J. Boling		LCdr.	J. Henson	⑧	ADRC	J. Fowlkes
Ltjg.	C. Garnett		Ltjg.	B. Kennedy		LCdr.	C. Smith		ADR2	M. Finch
Ltjg.	R. Farwell		RD1	H. Hoff		LCdr.	G. Williams		ATN2	D. Basinger
ADR1	R. Vance		AT1	C. Howard		Lt.	J. Harmin		ADR1	F. Choquette
AO2	L. Edwards		AO3	C. Johnson		ADR1	Donahue		AT1	A. Cohen
AMS2	J. Hicks		AX3	W. Hammatt		AX2	Shepherd		ADJ1	C. Cox
AX3	Woempner		AE2	C. Ranta		AO3	McNicholas		ADR3	F. Jackson
AX2	Moeller		AM3	W. Stenger		ATR3	J. Brown		Maj.	J. Elliott
AX1	F. Oster		AD2	J. Ullrich					Col.	R. Calland
AE2	M. Wood		AT3	W. Wentz					Capt.	J. Sullivan
AX2	T. Lewis		AZ2	W. Williams						

Note:

① Commanding Officer, Patrol Squadron Forty (VP-40)
② Commanding Officer, Aircraft Ferry Squadron Thirty-Two (VRF-32)
③ Commander Seventh Fleet: Last *Market Time* patrol by a Navy SP-5B. Marked the closing of last combat seadrome and tender operation
④ Commander Philippine Air Patrol Group: Officially ended the last deployment of the Martin SP-5B seaplane
⑤ Commander Fleet Air San Diego: Last operational flight by a U.S. Navy seaplane squadron
⑥ Executive Officer, VP-40
⑦ March 30, 1967: VAdm. A. M. Shinn, Commander Naval Air Forces U.S. Pacific Fleet, flew VP-40's QE-2, BuNo 147926 with Cdr. H. E. Longino on tour of *Market Time* area.
⑧ Pilot in command of last U.S. Navy seaplane crew to achieve *Alpha Status*. VP-40, Crew Ten in BuNo 135533

Appendix G

Final Flight of a Martin SP-5B Seaplane

(NAS North Island to NAS Patuxent River)

Summary

Manufacturer:	**Martin Aircraft Company**			**Delivery Date:**	**May 17, 1956**
Navy BuNo:	**135533**	**Type:**	**SP-5B "Marlin"**	**Total Hours:**	**2,737**
First Delivery:	**VX-1, FASRON 102**			**Last Squadron:**	**VP-40, (QE-10A)**

- **Ceremony at NAS North Island, California:** July 6, 1968

Capt.	H. Tate, USN	Commanding Officer, VRF-32
Capt.	W. Shockey, USN	Commanding Officer, Naval Air Rework Facility
Capt.	J. Sullivan, USNR	Martin-Marietta Representative
RAdm.	C. Karaberis, USN	Commander Fleet Air San Diego

- **Flight:**

Departure From North Island:		July 8, 1968
Intermediate Stops:	NAS Corpus Christi, Tx.	July 8, 1968
	NAS Jacksonville, Fl.	July 10, 1968
Final Destination:	NAS Patuxent River, Md.	July 12, 1968

Ferry Squadron: AIRFERRON Thirty-Two

Crew:			
	Pilot:	LCdr.	D. McGarrigle
	Co-Pilot:	Capt.	H. Tate
	PAO:	Ltjg.	M. Blackburn
		ADRC	J. Fowlkes
		ADR1	F. Choquette
		ADJ1	C. Cox
		ATN2	D. Basinger
		ATN2	A. Cohen
		ADR3	F. Jackson
		Capt.	J. Sullivan, Martin Representative *

* North Island to Jacksonville only

Members, National Armed Forces Museum Advisory Board, On Flight

Col.	R. Calland, USMC Ret.
Maj.	J. Elliott, USMC Ret.

- **Ceremony at NAS Patuxent River, Maryland:** July 12, 1968

VAdm.	T. Connolly, USN	Deputy Chief of Naval Operations for Air
Capt.	H. Tate, USN	VRF-32
	J. Brown	Smithsonian Regent
	J. Rauth	President, Martin-Marietta Aerospace Corporation
Capt.	V. O'Neil, USN	Commanding Officer, NAS Patuxent River

- -

Originally destined for display at the proposed National Armed Forces Museum in Washington, D.C., BuNo 135533 was stored at NAS Patuxent River until being acquired by the National Museum of Naval Aviation, Pensacola, Florida for permanent display.

Post-War Squadron Designator and Tail Code Development

With the outbreak of World War Two, the Navy's inventory of seaplanes and amphibians grew almost exponentially. Born out of increased growth came a complex system of squadron designations and markings. Following the war, the Navy reduced their inflated war-time inventory by demobilizing and consolidating squadrons. The result left the squadron numbering system out of sequence, with no apparent order. On November 15, 1946, a directive was issued by which the Navy could differentiate between squadron type and aircraft class. As part of this directive, the letters G, Y, J, N, O, Q were eliminated from the identification system.

Presently classed as VPB (Patrol, Bombing) or VP (Patrol), the seaplane squadrons were redesignated to reflect squadron type and class of aircraft. Under the new directive, the two primary categories became VP-MS (Patrol, Medium Seaplane) and VP-AM (Patrol, Amphibious). Tail code markings were also affected by this directive which resulted in an entirely new unit identification system. In early January 1947, the newly assigned unit identification letters were ordered displayed on both sides of the vertical fin and rudder and upper right and lower left wing surfaces. The system was again simplified on September 1, 1948, whereby all Patrol Squadrons reverted to the former VP designation, regardless of aircraft type. Although this primary system of squadron coding would remain in effect indefinitely from this point forward, the tail code marking system would again see a change in 1958.

Although greatly reduced, the Navy's post-war complement of seaplanes were still a formidable part of their inventory, and many were assigned to inactive reserve squadrons throughout the country. With the out-break of the Korean War, several seaplane squadrons were activated and assigned responsibilities in the Pacific. As reserve squadrons were activated in the early 1950s, the single tail letter used to identify their reserve station was removed and replaced with the tail code of the air group to which they were newly assigned; however, in addition to displaying the new two-letter tail code, some squadron aircraft also retained their reserve letter designator, displaying it elsewhere.

On September 30, 1958, the Navy again changed the tail code designator system as a further means of standardization. Now assigned two letters according to *fleet location*, i.e., AirLant (Atlantic) or AirPac (Pacific), this explains why some squadrons were required to change existing codes to conform to the new directive and others were not.

The number of Navy squadrons (Fleet) flying Martin medium-range seaplanes reached a peak in the late 1950s with a total of eleven in service. From September 1950 through July 1956, fleet squadrons continued to operate both PBM and P5M aircraft. From that point forward to November 1967, squadrons operated only the P5M-1 and P5M-2.

On September 18, 1962, a joint Army-Navy-Air Force regulation was issued, establishing a uniform system of designating aircraft similar to one previously used by the Air Force. Complying with this regulation, the Navy P5M series reflected the following changes;

Pre-1962:	**P5M-1**	Post-1962:	**P-5A**	ASW Upgrade:	**SP-5A**
Pre-1962:	**P5M-2**	Post-1962:	**P-5B**	ASW Upgrade:	**SP-5B**

Post-War Seaplane Squadron Development
1946 to 1950

Pre 1946	Post 1946	Fleet Air Wing	Tail Code	1948 to 1950	Tail Code
VP-201	VP-MS-1	11	MA	Decommissioned	----
VP-22	VP-MS-2	14	SA	VP-42	SA
VP-28	VP-MS-3	10	BC	VP-43	BC
VP-204	VP-MS-4	3	CC	VP-44	CC
VP-205	VP-MS-5	5	EE	VP-45	EE
VP-32	VP-MS-6	18	BD	VP-46	BD
VP-27	VP-MS-7	2	BA	VP-47	BA
VP-208	VP-MS-8	11	MB	VP-48	MB
VP-19	VP-MS-9	5	EA	VP-49	EA
VP-74	VP-MS-10	3	CA	VP-40	CA
VP-21	VP-MS-11	1	AA	VP-41	AA
VP-53	VP-AM-1	4	DA	Decommissioned	----
VP-62	VP-AM-2	4	DB	VP-32	DB
VP-71	VP-AM-3	5	EB	VP-33	EB
VP-73	VP-AM-4	5	EC	VP-34	EC

Note: VP = Patrol MS = Medium Seaplane AM = Amphibious

PBM and P5M Squadron Tail Codes
1950 to 1967

Training:
VT-31: 3A
VP-30: LL
VP-31: RP
ATU-10: ME
ATU-700/501: CB

Fleet:
VP-34: EC
VP-40: CA - QE
VP-42: SA - RB
VP-44: CC - LM
VP-45: EE - LN
VP-46: BD - RC
VP-47: BA - RD
VP-48: SF
VP-49: EA - LP
VP-50: SE - SG
VP-56: EH - LQ

Reserve:
VP-731: Designated VP-48, 1953
VP-661: Designated VP-56, 1953
VP-892: Designated VP-50, 1953

Note: ① Second set of tail codes to right of squadron number reflect 09-30-58 directive
② Post-war seaplane squadrons not listed were decommissioned between 1949 and 1950

Tail Code Directive
1958 to 1967

Command Location	First Letter	Second Letter
ComNavAirLant:	(A) through (M)	(A) through (Z)
ComNavAirPac :	(N) through (Z)	(A) through (Z)

Paint Scheme (USN)
1950 to 1967

Location/Color	Pre 1955	1955 to 1959	1959 to 1964	1964 to 1967
Top:	G Sea Blue	SG Seaplane Gray	G Insignia White	G Insignia White
Bottom:	G Sea Blue	SG Seaplane Gray	SG Seaplane Gray	NS Light Gull Gray

Note: ① G = Gloss SG = Semi-Gloss NS = Non-Specular
② Date of change: 02-23-55, 02-12-59, 06-29-64
③ Change to Insignia White upper on all patrol aircraft provided solar-reflecting roof

Appendix J

U.S. Seaplane Tenders

U.S. Naval seaplane tenders were classified by two different designations and often performed multiple roles throughout their individual careers before being retired or stricken from active service. The following represents a partial listing of popular examples from each designation.

(AV)
Seaplane Tender

Hull Number	Name	Commissioning Date	Class	Total In Class
5	Albemarle	12-20-40	Curtiss	2
7	Currituck	06-06-44	Currituck	4
9	Pocomoke	04-28-40	Pocomoke	2
12	Pine Island	04-26-45	Currituck	-
13	Salisbury Sound	11-26-45	Currituck	-
14	Kenneth Whiting	05-08-44	Kenneth Whiting	4

Total Commissioned in AV Class: 17

(AVP)
Seaplane Tender, Small

Hull Number	Name	Commissioning Date	Class	Total In Class
38	Duxbury Bay	12-31-44	Barnegat	32
39	Gardiners Bay	02-11-45	Barnegat	-
40	Floyds Bay	03-25-45	Barnegat	-
41	Greenwich Bay	05-20-45	Barnegat	-
49	Orca	01-23-44	Barnegat	-

Total Commissioned in AVP Class: 48

Note:

① Last ComNavAirLant seaplane tender on active duty: USS *Greenwich Bay* 1964
② Last ComNavAirPac seaplane tender on active duty: USS *Currituck* 1967
③ Last seaplane tender to earn Battle "E" Award: USS *Currituck* 1967
④ Between 1921 and 1946, a total of sixty-five seaplane tenders were commissioned in seven classes

Currituck Class Tenders

Of the many types of Seaplane Tenders (AV) commissioned between 1921 and 1946, the four in the Currituck Class were the largest; *Pine Island, Salisbury Sound, Norton Sound* and *Currituck.* Commissioned in the final years of World War Two, the Currituck Class tenders saw service in one form or another around the world, and three continued on throughout the duration of the seaplane with the Navy. In August 1951, a new classification of Auxiliary ship was established, AVM (Guided Missiles Ship). The USS *Norton Sound* (AV-11) was the first to be reclassified as AVM-1. The remaining three tenders in class continued to operate in their primary role throughout the final years of the seaplane and were phased out in steps.

The first to be retired was the *Pine Island.* Following its last tour of duty supporting the Vietnam conflict, the *Pine Island* returned to the United States in June 1966 and was decommissioned a year later on June 16, 1967. The second of the large tenders to be retired was the *Salisbury Sound.* Working in concert with the *Pine Island* until relieving her in June 1966, the *Salisbury Sound* completed its last ten month deployment to South East Asia and returned to the U.S. in November 1966. Retired to Puget Sound Naval Shipyard, Bremerton, Washington, the "Sally Sound" was decommissioned less than three months later.

The last of the big tenders to retire was the *Currituck* itself, the namesake of the largest class of naval seaplane tenders built. Returning to the U.S. from the last seaplane-tender operation with the Navy in May 1967, the *Currituck* was removed from active service and decommissioned on October 31, 1967.

Specifications
Currituck Class Tenders

Length:	540 feet 5 inches
Beam:	69 feet 3 inches
Displacement:	15,100 tons
Draft:	22 feet 3 inches
Speed:	19 knots
Crew:	1,250

Typical armament on this class tender consisted of four, five-inch guns; twenty, forty-millimeter; and twenty, twenty-millimeter guns. A more than adequate complement of defensive armament given its primary role.

USS *Salisbury Sound* (AV-13) anchored in Holmes Harbor while operating with VP-47 in June 1963. (U.S. Navy photo)

About The Author

Bruce D. Barth served as a Navy Aircrewman with VP-40, at the peak of seaplane involvement in Vietnam, between 1965 and 1967. A veteran of more than 40 combat patrols and over 700 flight hours in the Martin SP-5B, he participated in the last seaplane-tender operation with the U.S. Navy at Cam Ranh Bay, South Vietnam in April 1967. Working as an aircraft electrician aboard the SP-5B, he flew the position of Radio Operator and Ordnanceman and shared in the last *Alpha* awards earned by crews flying the *Marlin* in Vietnam. He continued to fly with the commanding officer's crew until November 1967 when the seaplanes were retired from the Navy's inventory.

Transferred in 1968 to a carrier squadron in South East Asia, attached to the USS *Kearsarge*, CVS-33, he continued to fly as a Radar Operator aboard the Grumman E-1B (early warning aircraft) with VAW-111. At the time of his honorable discharge from the Navy in 1969, Barth had accumulated over 1,100 flight hours and completed three operational deployments to South East Asia in support of the Vietnam conflict.

An historian on post-war Martin seaplanes, Barth has published several articles on the history of the Navy's last seaplane squadron and on the P5M series flying boat. He has worked as an aviation consultant on Martin seaplanes for several publications and is an active member of the Mariner / Marlin Association, VP-40 Reunion Group and the Glenn L. Martin Aviation Museum.

N
NW
NE
W
E
SW
SE
S
ANY WHERE - ANYTIME
MARINER/MARLIN ASSN

Ordering Information

To order additional copies of **"The Martin P5M Marlin"** please complete the attached order form and mail to:

Marlin Publication
P.O. Box 654
Belmont, California
94002-0654

(Dealer Inquiries Welcome)

"The Martin P5M Marlin"
(Please Print)

NAME:
____

ADDRESS:
____

CITY: ____________________ STATE: __________ ZIP: __________

Number of copies: ________ at **$19.95** per copy____________

Shipping & Handling: $3.00 ____________

TOTAL ____________

Payment: Enclose Check or Money Order payable to: **BRUCE D. BARTH**

Mail To: **Marlin Publication, P.O. Box 654, Belmont, California 94002-0654**